Lecture Notes in Control and Information Sciences

Volume 489

This series reports new developments in the fields of control and information sciences—quickly, informally and at a high level. The type of material considered for publication includes:

1. Preliminary drafts of monographs and advanced textbooks
2. Lectures on a new field, or presenting a new angle on a classical field
3. Research reports
4. Reports of meetings, provided they are
   (a) of exceptional interest and
   (b) devoted to a specific topic. The timeliness of subject material is very important.

Indexed by EI-Compendex, SCOPUS, Ulrich's, MathSciNet, Current Index to Statistics, Current Mathematical Publications, Mathematical Reviews, IngentaConnect, MetaPress and Springerlink.

More information about this series at https://link.springer.com/bookseries/642

Hideaki Ishii · Quanyan Zhu
Editors

Security and Resilience of Control Systems

Theory and Applications

Springer

Editors
Hideaki Ishii
Department of Computer Science
Tokyo Institute of Technology
Yokohama, Kanagawa, Japan

Quanyan Zhu
Department of Electrical and Computer Engineering
New York University
Brooklyn, NY, USA

ISSN 0170-8643 ISSN 1610-7411 (electronic)
Lecture Notes in Control and Information Sciences
ISBN 978-3-030-83235-3 ISBN 978-3-030-83236-0 (eBook)
https://doi.org/10.1007/978-3-030-83236-0

Mathematics Subject Classification: 91A10, 93-02, 93A16, 93B70, 93D15

This Springer imprint is published by the registered company Springer Nature Switzerland AG
The registered company address is: Gewerbestrasse 11, 6330 Cham, Switzerland

Preface

Developments toward a smart society require extensive communication connectivities among different devices for realizing cyber-physical systems involving feedback control in critical social infrastructures such as power, transportation, and automation. Control systems in these domains have steadily become more complex and large scale through increase in the integration of cyber-components (computation and communication) with physical systems. These structural changes in control systems, however, have created new entry points for adversarial cyber-attacks. Such attacks on control systems are fundamentally different from the conventional ones on information technology; physical systems can be monitored and even manipulated by adversaries, which degrade the availability of the control systems, disrupt the communication protocols needed for coordinated control, and compromise the safety and performance of physical components.

Security and privacy of smart control systems pose an inherently multidisciplinary research challenge. Networks and systems security is needed to discover possible cyber-vulnerabilities and to develop secure protocols. The incentives, capabilities, and goals of adversaries must be understood and modeled through techniques including fault-detection methods, fault-tolerant control, game theory, and distributed algorithms. The control actions should be robust to the impact of possible attacks and enable recovery of system functionality when attacks do occur. The difficulty is that all of these requirements must be met without impacting the real-time critical operation of the control systems.

This book presents a set of chapters that introduce various topics on novel approaches toward enhancing cyber-physical measures to increase security and resilience levels in control systems. Different from conventional security measures based on information technology, the unifying theme of these approaches lies in the utilization of the knowledge and models of the physical systems. It will be demonstrated that attacks in different forms such as false data injections and denial-of-service can be extremely harmful and may not be detected unless the security measures exploit properties specific to physical models. This new critical discipline is interdisciplinary in nature, necessitating techniques from several fields including systems control, game theory, signal processing, and computer science. For the applications, certain emphasis is placed on those in power systems.

The idea of editing this book has arisen from the series of workshops on energy management systems that took place over the period between 2016 and 2019 held in Heidelberg, Tokyo, and Mumbai. These were mainly organized by the US National Science Foundation (NSF) and the Japan Science and Technology Agency (JST). While the workshops had a broader scope addressing current problems related to the penetration of renewable energies in power systems, they provided chances for researchers with more particular interests in cyber-physical security to meet. Moreover, pre-conference workshops were organized by the editors at the IEEE Conference on Decision and Control in 2018, 2019, and 2020. These workshops had a more clear focus on security, and the authors of this book could further interact and discuss recent advances in security and privacy methods for control systems.

The book consists of eight chapters and are divided into three parts: Part I corresponds to the first three chapters, which deal with resilient control and estimation techniques under different classes of cyber-attacks including false data injections and jamming attacks. Next, Part II composed of the next two chapters places more emphases on secure communication and privacy issues in control systems, where the system operators should prevent the control-related data being accessed by adversaries. While the chapters in the first two parts study control systems taking general forms, the final part, Part III, addresses security problems within specific application domains of Global Positioning Systems (GPS) for vehicles and power systems.

It is evident at this time that research on security and privacy problems for cyber-physical systems is growing very fast, partly due to the numerous security incidents occurring in the world. We hope that this book will serve both as a source to gain an overview on the current research activities in this area and as a reference providing important indications and motivations for future studies. We are grateful to Rui Kato for his editorial assistance.

While finalizing this book, we heard the sad news that Dr. Radhakisan Baheti passed away on March 9, 2021, from the COVID-19 virus. Kishan was the NSF ECCS Program Director and one of the main organizers of the NSF–JST workshop series mentioned above. He was an avid supporter of our efforts toward international collaborations, which eventually led us to this edited volume. We would like to dedicate this book to him to pay homage to his leadership, enthusiasm, and mentorship over the years.

Yokohama, Japan Hideaki Ishii
New York, USA Quanyan Zhu
April 2021

Contents

Part I
Resilient Control and Estimation Under Cyber-Attacks

Chapter 1
Secured Filters Based on Saturated Innovations

Xingkang He, Xiaoqiang Ren, Henrik Sandberg, and Karl H. Johansson

Abstract In this chapter, we study how to design secure centralized and distributed filters for linear time-invariant systems with bounded noise under false data injection attacks in sensor networks. An adversary is able to compromise a subset of sensors and manipulate the measurements arbitrarily. We provide two motivating examples for this problem setup from smart buildings and autonomous vehicles. Then we design a centralized filter based on a saturation method, which gives a small gain if the innovation is large enough, indicating the high likelihood of compromised measurements. The estimation error of the secure centralized filter is proved to be asymptotically upper-bounded. Moreover, a secure two-time-scale distributed filter is obtained by modifying the centralized filter and employing an estimate consensus approach. Boundedness of the estimation error of the distributed filter is proved. Numerical simulations are provided in the end to show the usefulness of the two filters.

1.1 Introduction

Cyber-Physical Systems (CPSs) are systems with integrated computation, communication, and physical processes, which are controlled and monitored by computer-based algorithms [10]. During recent years, we have witnessed exponential develop-

---

X. He (✉) · H. Sandberg · K. H. Johansson
Division of Decision and Control Systems, School of Electrical Engineering and Computer Science, KTH Royal Institute of Technology, Stockholm, Sweden
e-mail: xingkang@kth.se

H. Sandberg
e-mail: hsan@kth.se

K. H. Johansson
e-mail: kallej@kth.se

X. Ren
School of Mechatronic Engineering and Automation, Shanghai University, Shanghai 200444, China
e-mail: xqren@shu.edu.cn

H. Ishii and Q. Zhu (eds.), *Security and Resilience of Control Systems*,
Lecture Notes in Control and Information Sciences 489,
https://doi.org/10.1007/978-3-030-83236-0_1

ment of CPSs in numerous applications, such as sensor networks, electric power grids, oil and natural gas distributions, process control, transportation systems, and so on [8]. Many of such systems are deployed in critical infrastructure, which have tremendous influence on our daily life. However, due to the CPS vulnerability induced by high integration of large-scale computer networks and complex physical processes, the security of CPSs becomes more and more important. Thus, designing secure and resilient solutions to improve the robustness of CPSs in malicious environments is increasingly drawing attention.

In a CPS, all kinds of sensors can be deployed to obtain system measurements, which can be used to estimate the system state or to design output feedback signals for the control of physical processes. In a malicious environment, an adversary can compromise measurements of some sensors by False Data Injection (FDI) attacks. Then, the observer or controller based on the compromised measurements will be unreliable, and even bring tremendous damage to the whole system. Thus, in order to mitigate the influence of attackers, some detection and identification schemes have been developed to monitor whether attacks exist in systems and if so how to identify the attack signals inserted into the systems. In [18], a study on attack detection and identification for CPSs was given, and meanwhile the design methods and analysis techniques for centralized and distributed monitors were investigated. A joint attack detection and state estimation approach was proposed by [4] in a Bayesian framework. In centralized frameworks where data centers exist, in order to obtain attack-resilient state estimates, some state estimators or observers were developed based on optimization techniques [3, 5, 16, 17, 20, 21], recursive implementation [15], and probabilistic approach [19]. In contrast, distributed frameworks do not require data centers, so they own advantages in quite a few aspects, such as structure robustness, scalability, and parallel processing. Due to the structural difference between centralized and distributed frameworks, it is usually difficult to extend centralized approaches into their distributed versions. In the study of distributed state estimation under Byzantine sensor attacks, observer-based methods were proposed for networks with sufficient structural robustness [14]. A distributed observer with an attack detection layer was given to deal with a class of bias attacks [2]. Distributed parameter estimation under FDI sensor attacks was studied in [1]. A distributed median solver-based observer was proposed for continuous-time systems [11].

In this chapter, we study the secure filtering problem for Linear Time-Invariant (LTI) systems with bounded noise under FDI sensor attacks. An adversary is able to compromise a subset of sensors and manipulate their measurements arbitrarily. We first provide two examples to motivate the problem. To mitigate the adversary's influence on estimation performance, we design a novel centralized filter by saturating the measurement innovation, which gives a small gain if the innovation is large enough, indicating the high likelihood of attacked measurements. The estimation error of the secure centralized filter is proved to be asymptotically upper-bounded. Moreover, we extend the centralized filter to its distributed version by employing an estimate consensus approach in the communication of neighboring sensors. The boundedness of the estimation error of the distributed filter is studied as well. The developed filters can be utilized to improve the resilience and robustness of cyber-physical systems under potential sensor attacks. The design of the distributed filter is based on [7], and the results on the centralized secure filter are novel.

The remainder of the chapter is organized as follows: Two motivating examples of secure filtering problems are provided in Sect. 1.2. Section 1.3 is on problem formulation. The algorithm design and performance analysis of a secure centralized filter and a distributed filter are given in Sects. 1.4 and 1.5, respectively. Numerical simulation results are provided in Sect. 1.6. Section 1.7 concludes this chapter.

Notation: $\mathbb{R}^{n\times m}$ is the set of real matrices with n rows and m columns. $\mathbb{R}^n$ is the set of n-dimensional real vectors. I_n stands for the n-dimensional square identity matrix. $\mathbf{1}_N$ stands for the N-dimensional vector with all elements being one. The superscript "T" represents the transpose. diag$\{\cdot\}$ represents the diagonalization operator. $A \otimes B$ is the Kronecker product of A and B. $\|x\|$ is the 2-norm of a vector x. $\|A\|$ is the induced 2-norm, i.e., $\|A\| = \sup_{x\neq 0} \frac{\|Ax\|}{\|x\|}$. $\lambda_{\min}(A)$, $\lambda_2(A)$, and $\lambda_{\max}(A)$ are the minimal, the second minimal, and the maximal eigenvalues of a symmetric real-valued matrix A, respectively. $|\Gamma|$ is the cardinality of the set Γ. $\mathbb{I}_{a\in\mathscr{A}}$ is an indicator function, which equals 1 if $a \in \mathscr{A}$, otherwise 0.

1.2 Motivating Examples

In this section, we provide two motivating examples to discuss the relevance of secure filtering problems under sensor attacks.

1.2.1 Security of Smart Buildings

Smart buildings are designed to automatically control building operations, such as lighting, heating, and air conditioning. To collect the related data of the buildings and monitor the change of the living/working environment, a lot of heterogeneous sensors are usually deployed in different building zones. The high integration of the connected sensors provides timely and effective reflection for emergencies like fire accident and gas leakage. However, once an adversary compromises the measurements of some sensors in a smart building by injecting faulty signals and the building system still uses the data normally, the building could face emergent events (e.g., fire accident). Therefore, to reinforce the security of smart buildings, it is necessary to design protection schemes and resilient strategies for the scenarios where some sensors are under attack.

The accurate estimation of people flow and room occupancy plays an important role in the decision-making of smart buildings, such as the emergency evacuation, heating, ventilation and air conditioning control, and elevator dispatching. We take the building occupancy in [12] as an example in the following. The goal of occupancy estimation is to count the number of people in each room of a building in real time.

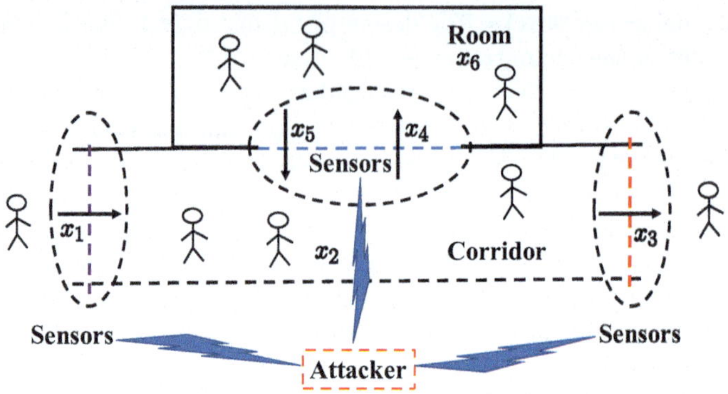

Fig. 1.1 A room is connected to a corridor through a door with sensors (e.g., beam sensors and cameras) which can monitor the number of people passing the door. The corridor has an entrance and an exit, and both of them are equipped with sensors to monitor how many people go through the doors. Some sensors could be attacked by an adversary in the sense that their measurements can be manipulated

The number is changing with the people flow through different rooms of the building. The situation is illustrated in Fig. 1.1, in which the numbers of people in a room and a corridor are to be estimated. For the situation in Fig. 1.1, we denote

- $x_1(t)$: Number of people moving into the corridor at time t;
- $x_2(t)$: Number of people in the corridor at time t;
- $x_3(t)$: Number of people moving out of the corridor at time t;
- $x_4(t)$: Number of people moving into the room at time t;
- $x_5(t)$: Number of people moving out of the room at time t;
- $x_6(t)$: Number of people in the room at time t.

According to [12], the system dynamics can be written as follows:

$$\begin{bmatrix} x_1(t+1) \\ x_2(t+1) \\ x_3(t+1) \\ x_4(t+1) \\ x_5(t+1) \\ x_6(t+1) \end{bmatrix} = \begin{bmatrix} \alpha_1 & 0 & 0 & 0 & 0 & 0 \\ 1 & 1 & -1 & -1 & 1 & 0 \\ 0 & \alpha_2 & 0 & 0 & 0 & 0 \\ 0 & \alpha_3 & 0 & 0 & 0 & 0 \\ 0 & 0 & 0 & 0 & 0 & \alpha_4 \\ 0 & 0 & 0 & 1 & -1 & 1 \end{bmatrix} \begin{bmatrix} x_1(t) \\ x_2(t) \\ x_3(t) \\ x_4(t) \\ x_5(t) \\ x_6(t) \end{bmatrix} + w(t),$$

where $w(t) \in \mathbb{R}^6$ stands for the uncertainty in the dynamics, and $\alpha_i \in [0, 1]$, $i = 1, 2, 3, 4$, are independent of the states but related to specific situations. The parameters α_i can be learned offline with identification methods.

There are groups of sensors (e.g., cameras and beam sensors) deployed at the doors, subject to

$$\begin{aligned} y_{1,i_1}(t) &= x_1(t) + a_{1,i_1}(t) + v_{1,i_1}(t), \quad i_1 = 1, 2, \ldots, N_1, \\ y_{2,i_2}(t) &= x_3(t) + a_{2,i_2}(t) + v_{2,i_2}(t), \quad i_2 = 1, 2, \ldots, N_2, \\ y_{3,i_3}(t) &= x_4(t) + a_{3,i_3}(t) + v_{3,i_3}(t), \quad i_3 = 1, 2, \ldots, N_3, \\ y_{4,i_4}(t) &= x_5(t) + a_{4,i_4}(t) + v_{4,i_4}(t), \quad i_4 = 1, 2, \ldots, N_4, \end{aligned} \tag{1.1}$$

where N_j stands for the sensor number at door j, $v_{j,i_j}(t)$ the measurement noise, and $a_{j,i_j}(t)$ the injected attack signal, $j = 1, 2, 3, 4$.

If the measurements in (1.1) are attack-free (i.e., $a_{j,i_j}(t) = 0$, $j = 1, 2, 3, 4$), existing robust filters (e.g., robust Kalman filter [12]) can be used to estimate the system state, i.e., the numbers of people in/out the room and corridor. However, if some sensors are under attack in the sense that their measurements are compromised, the estimation performance of the filters could be seriously degraded. Thus, in order to obtain accurate state estimates, it is quite necessary to design secure filters to mitigate the influence of the adversary.

1.2.2 Security of Autonomous Vehicles

Autonomous vehicles are designed to run on roads by timely identifying surroundings and detecting the relative positions between vehicles and objects nearby. This process relies on numerous sensors including GPS receiver, camera, radar, and lidar. Currently, the safety issue is one of the main problems affecting the extensive implementation of autonomous vehicles. Besides the inherent technical problems, potential threats from adversaries bring enormous challenges to driving safety. Due to the high digitalization of vehicle systems, the decision-making of vehicles seriously depends on sensor measurements, which are usually vulnerable under malicious environments. For example, considering an autonomous vehicle with a fast-driving speed, if an adversary successfully compromises some vehicle sensors such that the people in front of the vehicle cannot be timely detected, a serious accident could not be avoided.

As one of the promising applications of autonomous vehicles, vehicle platooning utilizes a formation control strategy such that a group of vehicles can run in a line by keeping the same desired speed and fixed distance between two nearest vehicles. In the development of intelligent transportation, vehicle platooning plays an important role since it enjoys high efficiency in energy saving and leads to improvement of the traffic flow. In the following, we take the example studied in [6] when the GPS data of one unknown vehicle is compromised (e.g., the GPS receiver is spoofed). The situation is illustrated in Fig. 1.2.

Assume each vehicle $i \in \{1, 2, \ldots, N\}$ satisfies the following dynamics:

$$x_i(t+1) = A_0 x_i(t) + \begin{bmatrix} 0 \\ T u_i(t) \end{bmatrix} + n_i(t), \tag{1.2}$$

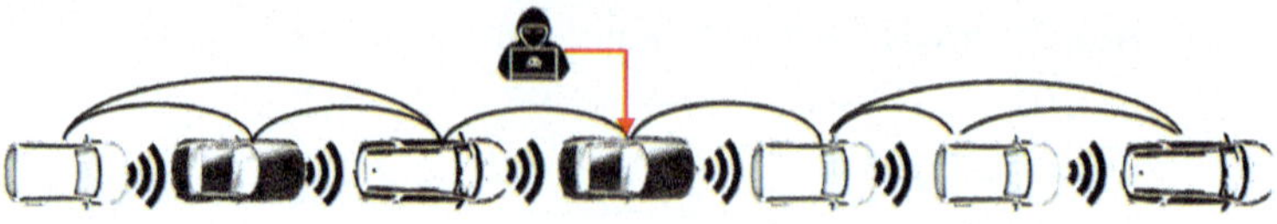

Fig. 1.2 A group of autonomous vehicles is running in a line. Each vehicle can observe the relative state between itself and its front vehicles through some secured sensors (e.g., camera or radar), and it can communicate with its neighboring vehicles through a wireless network (e.g., WiFi and 5G). Moreover, each vehicle can obtain its current position and velocity through a GPS receiver. In this case, the GPS receiver of vehicle 4 is attacked

where $x_i(t) = [s_i(t), v_i(t)]^\top \in \mathbb{R}^2$ is the state consisting of position $s_i(t)$ and velocity $v_i(t)$, $n_i(t) \in \mathbb{R}^2$ the process noise, $u_i(t) \in \mathbb{R}$ the acceleration input, $A_0 = \begin{bmatrix} 1 & T \\ 0 & 1 \end{bmatrix}$, and T the time step. Vehicle i obtains its position and velocity measurements from a GPS receiver which could be under attack:

$$y_i(t) = x_i(t) + a_i(t) + d_i(t), \tag{1.3}$$

where $y_i(t) \in \mathbb{R}^2$ and $d_i(t) \in \mathbb{R}^2$ are the measurement and measurement noise. The vector $a_i(t) \in \mathbb{R}^2$ is the attack signal injected by an adversary.

Note that if the system is attack-free, i.e., the GPS data is secure, each vehicle can use its own GPS data to estimate its position and velocity by some traditional methods (e.g., Kalman filter). Then, some formation control strategies can be used to achieve vehicle platooning based on the state estimates. However, once the measured GPS data is compromised as in (1.3), the measurements could lead to serious degradation in estimation performance, which would make the platoon fail and lead to the crash of vehicles.

In the following sections, we will formulate the problem and design-resilient filters in both centralized and distributed manners to estimate the states of LTI systems with potentially compromised measurements. The developed filters can be directly used in the first example on the occupancy estimation of smart buildings given the parameters α_i. In the second example, based on the estimates provided by our filters, existing distributed control strategies can be applied to achieve secure vehicle platooning [6].

1.3 Problem Formulation

In this section, we provide the system model, the adversary model, and the problems of interest.

1.3.1 System Model

Consider the following plant observed by N sensors:

$$\begin{aligned} x(t+1) &= Ax(t) + w(t), \\ y_i(t) &= C_i x(t) + v_i(t) + a_i(t), \quad i = 1, \ldots, N, \end{aligned} \tag{1.4}$$

where $x(t) \in \mathbb{R}^n$ is the system state, $w(t) \in \mathbb{R}^n$ the process noise, $y_i(t) \in \mathbb{R}$ the measurement of sensor i, $v_i(t) \in \mathbb{R}$ the measurement noise, and $a_i(t) \in \mathbb{R}$ the attack signal inserted by an adversary, all at time t. Moreover, $A \in \mathbb{R}^{n\times n}$ is the system state transition matrix, and $C_i \in \mathbb{R}^{1\times n}$ is the measurement vector of sensor i.

As seen, we assume the measurements of each sensor are scalar. Otherwise, one can easily extend the methods in this chapter to deal with the case with non-scalar measurements. The sensors studied in this chapter are smart sensors, meaning that these sensors have some limited capability in data storage, calculation, and communication. In a centralized framework, there is a data center which can collect all the measurements of each sensor to run a filter. However, in a distributed framework, each sensor can run a local filter and can communicate with other neighboring sensors in order to estimate the overall state collaboratively. The topologies of the centralized framework and the distributed framework under sensor attacks are illustrated in Figs. 1.3 and 1.4, respectively.

The following assumptions are needed in this chapter.

Assumption 1.1 The process noise, measurement noise, and the initial estimation error are all bounded, i.e., there exist nonnegative scalars b_w, b_v, and q_0 such that

$$\|w(t)\| \le b_w, \quad \left\|\hat{x}(0) - x(0)\right\| \le q_0, \quad \|v_i(t)\| \le b_v, \quad i = 1, \ldots, N,$$

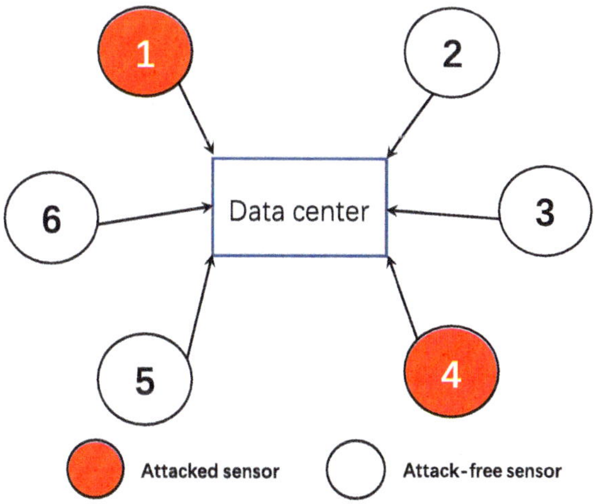

Fig. 1.3 A centralized framework with 6 sensors where 2 sensors are under attack

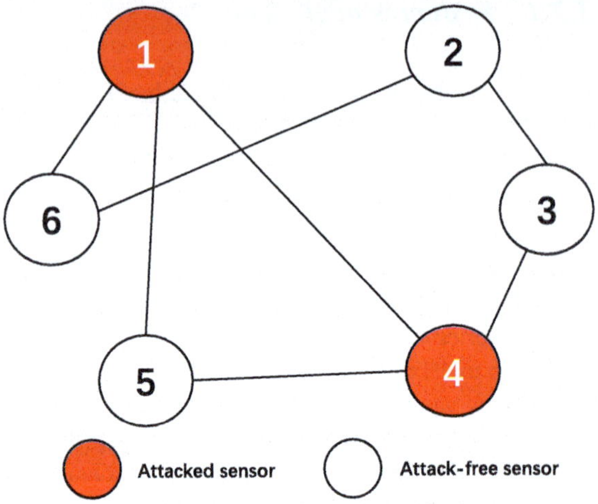

Fig. 1.4 A distributed framework with 6 sensors where 2 sensors are under attack

where $\hat{x}(0)$ is the estimate of $x(0)$ shared by all sensors and the data center.

Assumption 1.2 The measurement vector C_i is normalized, i.e., $\|C_i\| = 1$, $i = 1, \ldots, N$.

1.3.2 Attack Model

To deteriorate the filtering performance, an adversary is capable to compromise the measurements of some targeted sensors by injecting the attack signals $\{a_i(t)\}$ persistently. The knowledge of the adversary plays an important role in deteriorating the estimation performance. In the following, we assume the adversary has full knowledge of the model of the system (1.4), the network topology, and the filters of the data center, and all sensors. However, due to resource limitations, the adversary can only attack a subset of sensors.

Assumption 1.3 The adversary can arbitrarily manipulate measurements of up to s sensors among the N sensors, i.e.,

$$\begin{aligned} &a_i(t) \in \mathbb{R}, \quad i \in \mathscr{A}, \text{ with } |\mathscr{A}| \le s, \\ &a_i(t) = 0, \quad i \in \mathscr{N} = \{1, 2, \ldots, N\} \setminus \mathscr{A}, \quad \forall t \in \mathbb{N}, \end{aligned} \tag{1.5}$$

where $\mathscr{A}$ is the set of attacked sensors. Furthermore, the bound s is known to the system defender.

Note that both $\mathscr{A}$ and $\mathscr{N}$ are unknown to each sensor and the data center.

1.3.3 Problems of Interest

We mainly consider the following problems in this chapter.

(P1) How to design a secure centralized filter for a data fusion center? Under what conditions, is the boundedness of the estimation error of this filter guaranteed?

(P2) How to modify the centralized filter such that it is applicable in distributed frameworks?

1.4 Secure Centralized Filter Based on Saturated Innovations

In this section, we design a centralized filter and find conditions such that the estimation error of the filter is bounded.

We rewrite the system (1.4) in the following way:

$$\begin{aligned} x(t+1) &= Ax(t) + w(t), \\ y(t) &= Cx(t) + v(t) + a(t), \end{aligned} \tag{1.6}$$

where

$$y(t) = \begin{bmatrix} y_1(t) \\ y_2(t) \\ \vdots \\ y_N(t) \end{bmatrix}, \quad C = \begin{bmatrix} C_1 \\ C_2 \\ \vdots \\ C_N \end{bmatrix}, \quad v(t) = \begin{bmatrix} v_1(t) \\ v_2(t) \\ \vdots \\ v_N(t) \end{bmatrix}, \quad a(t) = \begin{bmatrix} a_1(t) \\ a_2(t) \\ \vdots \\ a_N(t) \end{bmatrix}. \tag{1.7}$$

The design of a centralized filter is to estimate the system state $x(t)$ based on the measurements $\{y(l), 0 \leq l \leq t\}$ and the messages received from neighbor sensors. We design the filter with two steps, namely for time update and measurement update. In the time update, a predictor is given based on the model information

$$\bar{x}(t) = A\hat{x}(t-1), \tag{1.8}$$

where $\hat{x}(t-1)$ is the estimate of $x(t-1)$. We denote the measurement innovation at time t by $\eta(t)$, where

$$\eta(t) = y(t) - C\bar{x}(t). \tag{1.9}$$

Let

$$\eta(t) := \begin{bmatrix} \eta_1(t) \\ \eta_2(t) \\ \vdots \\ \eta_N(t) \end{bmatrix}. \tag{1.10}$$

To utilize the measurement innovation $\eta_i(t)$, a saturation method is provided as follows. For each $i \in \{1, 2, \ldots, N\}$, calculate

Algorithm 1.1 Centralized Saturation-Based Filter (CSF):

Initialization: Initial estimate $\hat{x}(0)$, and saturation parameter β
Output: State estimates $\{\hat{x}(t)\}_{t=0,1,\ldots}$
for $t = 1, 2, \ldots$ **do**
  Time update: Data center carries out a state predictor
  $\bar{x}(t) = A\hat{x}(t-1)$.
  Measurement update: Data center uses the measurement $y(t)$ to update the estimate
  $\left[\eta_1(t)\ \eta_2(t)\ \ldots\ \eta_N(t)\right]^{\mathsf{T}} := \eta(t) = y(t) - C\bar{x}(t)$.
  for $i = 1, 2, \ldots, N$ **do**
    $k_i(t) = \min\{\frac{1}{N}, \frac{\beta}{N|\eta_i(t)|}\}$.
  end for
  $K(t) := \mathrm{diag}\{k_1(t), k_2(t), \ldots, k_N(t)\}$,
  $\hat{x}(t) = \bar{x}(t) + C^{\mathsf{T}}K(t)\eta(t)$.
end for

$$k_i(t) = \begin{cases} \frac{1}{N}, & \text{if } |\eta_i(t)| \le \beta, \\ \frac{\beta}{N|\eta_i(t)|}, & \text{otherwise,} \end{cases} \tag{1.11}$$

where $\beta > 0$ is a saturation parameter to be designed. Denote

$$K(t) := \mathrm{diag}\{k_1(t), k_2(t), \ldots, k_N(t)\},$$

then the measurement update is

$$\hat{x}(t) = \bar{x}(t) + C^{\mathsf{T}}K(t)\eta(t). \tag{1.12}$$

By (1.8)–(1.12), we obtain the Centralized Saturation-based Filter (CSF) in Algorithm 1.

Different from traditional filters (e.g., Kalman filter), the gain $k_i(t)$ in Algorithm 1 is related to the innovation of sensor i (i.e., $y_i(t) - C_iA\hat{x}(t-1)$). The design of $k_i(t)$ in (1.11) makes sense, since if the innovation is very large, the measurement $y_i(t)$ is more likely to be compromised. By the designed gain $k_i(t)$, we have $|k_i(t)(y_i(t) - C_iA\hat{x}(t-1))| \le \frac{\beta}{N}$, which ensures that the adversary has limited influence in the measurement update. The parameter β in (1.11) reflects the usage trade-off between normal measurements and compromised measurements. If β is very large, many normal measurements will be used without scaling. However, it will give much space for the adversary to deteriorate the estimation performance. If β is very small, although most possible attack signals may be filtered, normal measurements may contribute little to the estimation performance. As a result, the filtering error of each sensor will probably be divergent for unstable systems. The design of β will be discussed later.

Since the filtering gains $\{k_i(t)\}$ are related to the state estimates and potentially compromised measurements, the common stability analysis approaches, such as Lyapunov methods, may not be directly utilized to analyze stability or boundedness of the estimation error. This is the main technical challenge in the study of Algorithm 1.

Next, we aim to find a condition under which the estimation error of Algorithm 1 is bounded. For a given set $\mathscr{I} \subset \{1, 2, \ldots, N\}$ with $|\mathscr{I}| \leq N$, denote $\mathscr{I}^c = \{1, 2, \ldots, N\} \setminus \mathscr{I}$, and

$$\begin{aligned} K_{\mathscr{I}}(t) &= \mathrm{diag}\{k_1(t)\mathbb{I}_{1\in\mathscr{I}}, k_2(t)\mathbb{I}_{2\in\mathscr{I}}, \ldots, k_N(t)\mathbb{I}_{N\in\mathscr{I}}\}, \\ K_{\mathscr{I}^c}(t) &= \mathrm{diag}\{k_1(t)\mathbb{I}_{1\in\mathscr{I}^c}, k_2(t)\mathbb{I}_{2\in\mathscr{I}^c}, \ldots, k_N(t)\mathbb{I}_{N\in\mathscr{I}^c}\}. \end{aligned} \tag{1.13}$$

It holds that $K(t) = K_{\mathscr{I}^c}(t) + K_{\mathscr{I}}(t)$. The following lemma provides an upper bound of the estimation error at each time.

Lemma 1.1 *Consider the system* (1.4) *and Algorithm 1. Under Assumptions 1.1–1.3, the estimation error* $e(t) = \hat{x}(t) - x(t)$ *satisfies*

$$\|e(t)\| \leq q(t),$$

where

$$\begin{aligned} q(t) &= (1 - k(t)\lambda_0)\,\|A\|\,q(t-1) + f(\beta, s), \quad q(0) = q_0, \\ k(t) &= \min\left\{\frac{1}{N}, \frac{\beta}{N(\|A\|\,q(t-1) + b_w + b_v)}\right\}, \\ f(\beta, s) &= \frac{N-s}{N}(b_w + b_v) + b_w + \frac{s\beta}{N}, \\ \lambda_0 &= \min_{\mathscr{I}\subset\{1,2,\ldots,N\}:|\mathscr{I}|=N-s} \lambda_{\min}\left(\sum_{i\in\mathscr{I}} C_i^\top C_i\right). \end{aligned} \tag{1.14}$$

Proof Under Assumption 1.3, there are at least $N - s$ attack-free sensors. Suppose $\mathscr{I}$ is the set of these $N - s$ sensors, i.e., $\mathscr{I} \subset \mathscr{A}^c$ with $|\mathscr{I}| = N - s$. By Algorithm 1 and denotations in (1.13), we have

$$
\begin{aligned}
e(t) &= Ae(t-1) + C^{\mathsf{T}}K(t)(y(t) - C\bar{x}(t)) - w(t-1) \\
&= Ae(t-1) + C^{\mathsf{T}}(K_{\mathscr{J}}(t) + K_{\mathscr{J}^c}(t))(y(t) - C\bar{x}(t)) - w(t-1) \\
&= (I - C^{\mathsf{T}}K_{\mathscr{J}}(t)C)Ae(t-1) + C^{\mathsf{T}}K_{\mathscr{J}}(t)(Cw(t-1) + v(t)) \\
&\quad + C^{\mathsf{T}}K_{\mathscr{J}^c}(t)(y(t) - C\bar{x}(t)) - w(t-1) \\
&= (I - \sum_{i\in\mathscr{J}} k_i(t)C_i^{\mathsf{T}}C_i)Ae(t-1) \\
&\quad + \sum_{i\in\mathscr{J}} k_i(t)C_i^{\mathsf{T}}(v_i(t) + C_iw(t-1)) \\
&\quad - w(t-1) + \sum_{i\in\mathscr{J}^c} k_i(t)C_i^{\mathsf{T}}\eta_i(t).
\end{aligned} \tag{1.15}
$$

By Assumptions 1.1–1.2, and $k_i(t) \le \frac{1}{N}$, we have

$$
\left\| \sum_{i\in\mathscr{J}} k_i(t)C_i^{\mathsf{T}}(v_i(t) + C_iw(t-1)) - w(t-1) \right\| \le \frac{N-s}{N}(b_v + b_w) + b_w. \tag{1.16}
$$

Since $|\mathscr{J}^c| = s$ and $\|C_i\| = 1$, it follows from (1.11) that

$$
\left\| \sum_{i\in\mathscr{J}^c} k_i(t)C_i^{\mathsf{T}}\eta_i(t) \right\| \le \frac{s\beta}{N}. \tag{1.17}
$$

By (1.16) and (1.17), taking the norm on both sides of (1.15) yields

$$
\|e(t)\| \le \left\| (I - \sum_{i\in\mathscr{J}} k_i(t)C_i^{\mathsf{T}}C_i) \right\| \|A\| \, \|e(t-1)\| + f(\beta, s). \tag{1.18}
$$

To prove $\|e(t)\| \le q(t)$, an inductive method is used. At the initial time, it holds trivially under Assumption 1.1. Suppose at time $t-1$, $\|e(t-1)\| \le q(t-1)$. Then for an attack-free sensor $i \in \mathscr{J}$, its innovation is $\eta_i(t) = C_iAe(t-1) + C_iw(t-1) + v_i(t)$. By $\|e(t-1)\| \le q(t-1)$ and Assumption 1.1, we have

$$
|\eta_i(t)| \le \|A\|\, q(t-1) + b_w + b_v.
$$

As a result, for $i \in \mathscr{J}$, the design of the gain $k_i(t)$ in (1.11) leads to

$$
k_i(t) \ge \min\left\{ \frac{1}{N}, \frac{\beta}{N(\|A\|\, q(t-1) + b_w + b_v)} \right\} =: k(t).
$$

Due to $\|C_i\| = 1$ and $k_i(t) \le \frac{1}{N}$,

$$0 \preceq \sum_{i \in \mathscr{J}} k_i(t) C_i^\mathsf{T} C_i \preceq I.$$

Then we have

$$\begin{aligned}\left\| I - \sum_{i \in \mathscr{J}} k_i(t) C_i^\mathsf{T} C_i \right\| &= \lambda_{\max} \left(I - \sum_{i \in \mathscr{J}} k_i(t) C_i^\mathsf{T} C_i \right) \\ &\leq \lambda_{\max} \left(I - k(t) \sum_{i \in \mathscr{J}} C_i^\mathsf{T} C_i \right) \\ &\leq 1 - k(t)\lambda_0.\end{aligned}$$

From (1.18), $\|e(t)\| \leq q(t)$ holds. □

The following theorem provides a sufficient condition such that the estimation error of Algorithm 1 is asymptotically upper-bounded.

Theorem 1.1 *Consider the system* (1.4) *and Algorithm 1. Under Assumptions 1.1–1.3, if*

$$(1 - k_1 \lambda_0) \|A\| q_0 + f(\beta, s) < q_0, \tag{1.19}$$

the estimation error is asymptotically upper-bounded, i.e.,

$$\limsup_{t \to \infty} \|e(t)\| \leq \frac{f(\beta, s)}{1 - \mu} < q_0, \tag{1.20}$$

where $\mu = (1 - k_1 \lambda_0) \|A\|$ *and* $k_1 = \min \left\{ \frac{1}{N}, \frac{\beta}{N(\|A\| q_0 + b_w + b_v)} \right\}$.

Proof Under (1.19), $\mu \in [0, 1)$, then $\frac{f(\beta, s)}{1-\mu} < q_0$. First, we prove $q(t) \leq q_0$ in the following. It holds trivially at time $t = 0$. Assume at time $t - 1 \geq 0$, $q(t-1) \leq q_0$. At time t, from (1.14) we have

$$k(t) \geq \min \left\{ \frac{1}{N}, \frac{\beta}{N(\|A\| q_0 + b_w + b_v)} \right\} = k_1.$$

Also, it holds that

$$\begin{aligned} q(t) &= (1 - k(t)\lambda_0) \|A\| q(t-1) + f(\beta, s) \\ &\leq (1 - k_1 \lambda_0) \|A\| q(0) + f(\beta, s) \\ &< q_0. \end{aligned}$$

Thus, $q(t) \leq q_0$. As a result, for $\forall t \geq 0$, $k(t) \geq k_1$. Then

$$q(t) = (1 - k(t)\lambda_0) \|A\| q(t-1) + f(\beta, s)$$
$$\leq \mu q(t-1) + f(\beta, s).$$

Next, we prove $\limsup_{t\to\infty} q(t) \leq \frac{f(\beta,s)}{1-\mu}$, which holds trivially if $\mu = 0$. For the case that $\mu \in (0, 1)$,

$$\limsup_{t\to\infty} q(t) = \limsup_{t\to\infty} \left(\mu^t q(0) + f(\beta, s) \sum_{j=0}^{t-1} \mu^j \right) = \frac{f(\beta, s)}{1-\mu}.$$

By Lemma 1.1, the conclusion holds. □

The condition (1.19) can be examined offline with global knowledge. To improve the estimation performance, one way for the system defender is to lower the bound in (1.20) by designing a proper parameter β offline based on global knowledge.

1.5 Secure Distributed Filter: A Consensus Method

In this section, we extend the centralized filter in Sect. 1.4 into a distributed filter. Before that, some graph preliminaries are provided.

1.5.1 Graph Preliminaries

We model the sensor network topology by an undirected graph $\mathcal{G} = (\mathcal{V}, \mathcal{E}, \mathcal{A})$, where $\mathcal{V} = \{1, 2, \ldots, N\}$ is the set of nodes, $\mathcal{E} \subseteq \mathcal{V} \times \mathcal{V}$ the set of edges, and $\mathbb{A} = [a_{i,j}]_{i,j=1}^N$ the adjacency matrix which is symmetric. If there is an edge $(i, j) \in \mathcal{E}$, then $a_{i,j} = 1$, otherwise, $a_{i,j} = 0$. If $(i, j) \in \mathcal{E}$, nodes i and j can exchange information with each other. Then node j is called a neighbor of node i, and vice versa. Let the neighbor set of sensor i be $\mathcal{N}_i := \{j \in \mathcal{V} | (i, j) \in \mathcal{E}\}$. The matrix $\mathcal{L} = \mathcal{D} - \mathbb{A}$ is called Laplacian matrix, where $\mathcal{D}$ is the degree matrix consisting of numbers of neighboring nodes, i.e., $\mathcal{D} = \text{diag}\{|\mathcal{N}_1|, |\mathcal{N}_2|, \ldots, |\mathcal{N}_N|\}$. The graph $\mathcal{G}$ is connected if for any pair of nodes i_1, i_l, there exists a path from i_1 to i_l consisting of edges $(i_1, i_2), (i_2, i_3), \ldots, (i_{l-1}, i_l)$. On the connectivity of $\mathcal{G}$, we have the following proposition (e.g., [13]).

Proposition 1.1 *The undirected graph $\mathcal{G}$ is connected if and only if $\lambda_2(\mathcal{L}) > 0$.*

On the topology of the sensor network, the following assumption is needed.

Assumption 1.4 The communication graph $\mathcal{G}$ is connected.

Although [14] studied a wider range of attacks than in this chapter, it required some robustness of communication graph. In contrast, the connectivity of Assumption 1.4 is a standard condition for distributed estimation. If the graph $\mathcal{G}$ is not connected, the problem can be studied for the connected subgraphs separately.

A typical distributed filtering problem is to design an online filter for each sensor i to estimate the system state $x(t)$ by employing the known local noisy measurements $\{y_i(l)\}_{l=1}^{t}$ and the messages received from neighboring sensors. However, if measurements of some sensors are compromised by an adversary, the measurement quality may be tremendously affected, which will bring big challenges in the design and analysis of distributed filtering algorithms.

1.5.2 Distributed Filter Design

Next, we design a distributed filter with three steps for, namely time update, measurement update, and estimate consensus. The reason for estimate consensus is to update the estimate of each sensor with the global measurement information, which helps to improve the estimation performance when the sensor is under attack or in the absence of local observability as illustrated in the motivating examples.

Algorithm 1.2 Distributed Saturation-Based Filter (DSF):

Initialization: Initial estimate $\hat{x}_i(0)$, consensus parameter α, saturation parameter β, consensus parameter L.
Output: State estimates $\{\hat{x}_i(t)\}_{t=0,1,\ldots}$
for $t = 1, 2, \ldots$ **do**
  Time update: For each sensor i
  $\bar{x}_i(t) = A\hat{x}_i(t-1)$,
  Measurement update: For each sensor i
  $\tilde{x}_i(t) = \bar{x}_i(t) + k_i(t)C_i^{\mathrm{T}}(y_i(t) - C_i\bar{x}_i(t))$.
  $k_i(t) = \min\{1, \frac{\beta}{\|y_i(t)-C_i\bar{x}_i(t)\|}\}$.
  Estimate consensus: For each sensor i, $\hat{x}_{i,0}(t) = \tilde{x}_i(t)$
  for $l = 1, \ldots, L$ **do**
    Sensor i receives $\hat{x}_{j,l-1}(t)$ from neighbor sensor j,
    $\hat{x}_{i,l}(t) = \hat{x}_{i,l-1}(t) - \alpha\sum_{j\in\mathcal{N}_i}(\hat{x}_{i,l-1}(t) - \hat{x}_{j,l-1}(t))$.
  end for
  Let $\hat{x}_i(t) = \hat{x}_{i,L}(t)$.
end for

In the time update, for sensor i, we use the model information for prediction

$$\bar{x}_i(t) = A\hat{x}_i(t-1), \tag{1.21}$$

where $\hat{x}_i(t-1)$ is the estimate of $x(t-1)$ by sensor i.

In the local measurement update, similar to the centralized method,

$$\tilde{x}_i(t) = \bar{x}_i(t) + k_i(t)C_i^{\mathrm{T}}(y_i(t) - C_i\bar{x}_i(t)), \tag{1.22}$$

where

$$k_i(t) = \begin{cases} 1, & \text{if } |y_i(t) - C_i\bar{x}_i(t)| \leq \beta, \\ \frac{\beta}{|y_i(t) - C_i\bar{x}_i(t)|}, & \text{otherwise}. \end{cases} \tag{1.23}$$

where $\beta > 0$ is a saturation parameter to be determined.

In the estimate consensus, we consider a two-time-scale scheme, where each sensor is able to communicate with its neighboring sensors for $L \geq 1$ times between two measurement updates. During each communication, sensor j transmits its estimate $\hat{x}_{j,l-1}(t)$ to its neighbors, $l = 1, \ldots, L$. For each l, calculate

$$\hat{x}_{i,l}(t) = \hat{x}_{i,l-1}(t) - \alpha \sum_{j \in \mathcal{N}_i} (\hat{x}_{i,l-1}(t) - \hat{x}_{j,l-1}(t)), \tag{1.24}$$

with $\hat{x}_{i,0}(t) = \bar{x}_i(t)$, $\hat{x}_i(t) = \hat{x}_{i,L}(t)$, and α is a consensus parameter to be designed.

Summarizing (1.21)–(1.24), we obtain the Distributed Saturation-based Filter (DSF) in Algorithm 2.

The consensus term $\alpha \sum_{j \in \mathcal{N}_i} (\hat{x}_{i,l-1}(t) - \hat{x}_{j,l-1}(t))$ is to make the estimates of sensors close to each other. It can be proved that if the consensus step L goes to infinity and the parameter α is properly designed, the estimates $\{\hat{x}_i(t)\}_{i=1}^N$ will converge to the same vector. However, the consensus step L is not necessary to approximate infinity in our approach. The requirement of the step L and the design of α are studied in the next section.

Different from the centralized gain design in Algorithm 1, where

$$K(t) = \mathrm{diag}\{k_i(t)\}_{i=1}^N, \text{ with } k_i(t) \leq \frac{1}{N},$$

the filter gain $k_i(t)$ in Algorithm 2 is a scalar and $k_i(t) \leq 1$. Because the centralized algorithm can use the measurements of all sensors at each time, a large gain could lead to instability of the estimation error.

1.5.3 *Performance Analysis*

We denote

$$\begin{aligned} X(t) &= \mathbf{1}_N \otimes x(t), \\ \hat{X}(t) &= \left[\hat{x}_1^{\mathsf{T}}(t), \ldots, \hat{x}_N^{\mathsf{T}}(t)\right]^{\mathsf{T}}, \\ E(t) &= \hat{X}(t) - X(t), \\ \bar{C} &= \mathrm{diag}\{C_1, \ldots, C_N\}, \\ K(t) &= \mathrm{diag}\{k_1(t), \ldots, k_N(t)\}, \\ P_{Nn} &= \frac{1}{N}(\mathbf{1}_N \otimes I_n)(\mathbf{1}_N \otimes I_n)^{\mathsf{T}}. \end{aligned} \tag{1.25}$$

We obtain the compact form of recursive state estimates of DSF in the following:

$$\hat{X}(t) = (I_{Nn} - \alpha(\mathscr{L} \otimes I_n))^L \Big[(I_N \otimes A)\hat{X}(t-1) + \bar{C}^{\mathsf{T}} K(t)(y(t) - \bar{C}(I_N \otimes A)\hat{X}(t-1))\Big], \tag{1.26}$$

where $y(t)$ is introduced in (1.7).

Before providing the main result on the boundedness of the estimation error (i.e., $e_i(t) = \hat{x}_i(t) - x(t)$) of Algorithm 2, we give the proof idea as follows: We can split the estimation error into two parts

$$e_i(t) = \bar{e}_i(t) + \tilde{e}(t), \tag{1.27}$$

where

$$\bar{e}_i(t) = \hat{x}_i(t) - \frac{1}{N}\sum_{i=1}^{N}\hat{x}_i(t), \quad \tilde{e}(t) = \frac{1}{N}\sum_{i=1}^{N}\hat{x}_i(t) - x(t). \tag{1.28}$$

The vector $\bar{e}_i(t)$ indicates the difference between the estimates of all sensors. Given a proper designed α, when L goes to infinity, $\bar{e}_i(t)$ will tend to zero. The vector $\tilde{e}(t)$ stands for the networked tracking error, which shows the estimation error of the averaged sensor estimate. To keep $e_i(t)$ bounded, in the following, we will study the boundedness of $\bar{e}_i(t)$ and $\tilde{e}(t)$, respectively. Some useful lemmas are provided and proved.

Lemma 1.2 *The following equation holds:*

$$(I_{Nn} - \alpha(\mathscr{L} \otimes I_n) - P_{Nn})^L M_t = (I_{Nn} - \alpha(\mathscr{L} \otimes I_n))^L M_t,$$

where $M_t = (I_N \otimes A)(\hat{X}(t) - \mathbf{1}_N \otimes (\frac{1}{N}\sum_{i=1}^{N}\hat{x}_i(t)))$.

Proof Recall the denotations in (1.25), then we have

$$P_{Nn}M_t = (I_N \otimes A)P_{Nn}(\hat{X}(t) - \mathbf{1}_N \otimes (\frac{1}{N}\sum_{i=1}^{N}\hat{x}_i(t))) = 0.$$

Thus,

$$(I_{Nn} - \alpha(\mathscr{L} \otimes I_n) - P_{Nn})\, M_t = (I_{Nn} - \alpha(\mathscr{L} \otimes I_n))\, M_t.$$

Due to

$$
\begin{aligned}
&(I_{Nn} - \alpha(\mathscr{L} \otimes I_n) - P_{Nn})\,(I_{Nn} - \alpha(\mathscr{L} \otimes I_n)) \\
&\quad = (I_{Nn} - \alpha(\mathscr{L} \otimes I_n))\,(I_{Nn} - \alpha(\mathscr{L} \otimes I_n) - P_{Nn}),
\end{aligned}
$$

by the law of commutation and recursively applying the above operation for L times, the conclusion holds. □

The following lemma provides a design for the consensus parameter α, which is related to the graph topology.

Lemma 1.3 *Under Assumption 1.4, the following relation holds:*

$$
\min_{\alpha} \|I_{Nn} - \alpha(\mathscr{L} \otimes I_n) - P_{Nn}\| = \frac{\lambda_{max}(\mathscr{L}) - \lambda_2(\mathscr{L})}{\lambda_{max}(\mathscr{L}) + \lambda_2(\mathscr{L})} < 1,
$$

with the optimal solution $\alpha^* = \frac{2}{\lambda_2(\mathscr{L})+\lambda_{max}(\mathscr{L})}$.

Proof Since the eigenvalues of $I_{Nn} - \alpha(\mathscr{L} \otimes I_n) - P_{Nn}$ are zero and $1 - \alpha\lambda_i(\mathscr{L})$, $i = 2, \ldots, N$, with each repeated for n times, we have

$$
\|I_{Nn} - \alpha(\mathscr{L} \otimes I_n) - P_{Nn}\| = \max\{|1 - \alpha\lambda_2(\mathscr{L})|, |1 - \alpha\lambda_{max}(\mathscr{L})|\}.
$$

It can be easily proved that if

$$
1 - \alpha\lambda_2(\mathscr{L}) = \alpha\lambda_{max}(\mathscr{L}) - 1,
$$

$\|I_{Nn} - \alpha(\mathscr{L} \otimes I_n) - P_{Nn}\|$ is minimized. Thus, solving $1 - \alpha\lambda_2(\mathscr{L}) = \alpha\lambda_{max}(\mathscr{L}) - 1$ yields $\alpha = \frac{2}{\lambda_2(\mathscr{L})+\lambda_{max}(\mathscr{L})}$, which leads to

$$
\|I_{Nn} - \alpha(\mathscr{L} \otimes I_n) - P_{Nn}\| = \frac{\lambda_{max}(\mathscr{L}) - \lambda_2(\mathscr{L})}{\lambda_{max}(\mathscr{L}) + \lambda_2(\mathscr{L})}.
$$

Under Assumption 1.4 and Proposition 1.1, it holds that $\lambda_2(\mathscr{L}) > 0$, which guarantees $\frac{\lambda_{max}(\mathscr{L})-\lambda_2(\mathscr{L})}{\lambda_{max}(\mathscr{L})+\lambda_2(\mathscr{L})} < 1$. □

The following lemma studies the influence of consensus step L on the measurement innovation part in (1.26).

Lemma 1.4 *Let* $\alpha = \frac{2}{\lambda_2(\mathscr{L})+\lambda_{max}(\mathscr{L})}$. *Then*

$$
\left\|(I_{Nn} - P_{Nn})\,(I_{Nn} - \alpha(\mathscr{L} \otimes I_n))^L\, \bar{y}(t)\right\| \le \sqrt{N}\gamma^L\beta,
$$

where

$$
\begin{aligned}
\gamma &= \frac{\lambda_{max}(\mathscr{L}) - \lambda_2(\mathscr{L})}{\lambda_{max}(\mathscr{L}) + \lambda_2(\mathscr{L})}, \\
\bar{y}(t) &= \bar{C}^{\mathsf{T}}\bar{K}(t)(y(t) - \bar{C}(I_N \otimes A)\hat{X}(t-1)).
\end{aligned}
\tag{1.29}
$$

Proof Consider

$$\begin{aligned}
&\left\|(I_{Nn} - P_{Nn})\,(I_{Nn} - \alpha(\mathscr{L} \otimes I_n))^L\, \bar{y}(t)\right\| \\
&= \left\|(I_{Nn} - \alpha(\mathscr{L} \otimes I_n))^L\,(I_{Nn} - P_{Nn})\bar{y}(t)\right\| \\
&\le (\max\{|1 - \alpha\lambda_2(\mathscr{L})|, |1 - \alpha\lambda_{max}(\mathscr{L})|\})^L\, \|(I_{Nn} - P_{Nn})\bar{y}(t)\| \\
&\le \left(\frac{\lambda_{max}(\mathscr{L}) - \lambda_2(\mathscr{L})}{\lambda_{max}(\mathscr{L}) + \lambda_2(\mathscr{L})}\right)^L \sqrt{N}\beta \\
&= \gamma^L \sqrt{N}\beta,
\end{aligned}$$

where the first inequality is obtained by Lemma 4.4 in [9] and by noting that $(I_{Nn} - P_{Nn})\bar{y}(t)$ belongs to the set orthogonal to the set of consensus states. □

Under Assumption 1.4, $\gamma \in [0, 1)$. Define $\gamma^{-1} = +\inf$ if $\gamma = 0$. In the following lemma, the boundedness of $\bar{e}_i(t)$ is studied.

Lemma 1.5 *Consider Algorithm 2, and let Assumptions 1.1–1.4 hold. By choosing* $\alpha = \frac{2}{\lambda_2(\mathscr{L}) + \lambda_{max}(\mathscr{L})}$, *and* $L > \frac{\ln\|A\|}{\ln \gamma^{-1}}$, *for* $t \ge 0$,

$$\|\bar{e}_i(t)\| \le g(t) := \sqrt{N}\beta\gamma^L \frac{1 - \|A\|^t \gamma^{Lt}}{1 - \|A\|\,\gamma^L}. \tag{1.30}$$

Proof Denote $\hat{x}_{avg}(t) = \frac{1}{N}\sum_{i=1}^{N} \hat{x}_i(t)$, $\bar{E}(t) = [\bar{e}_1(t)^\top, \ldots, \bar{e}_N(t)^\top]^\top$, where $\bar{e}_i(t)$ is defined in (1.28). According to (1.26) and the following properties

$$\begin{aligned}
P_{Nn}\hat{X}(t) &= \mathbf{1}_N \otimes \hat{x}_{avg}(t), \\
P_{Nn}\left(\mathbf{1}_N \otimes x_{avg}(t)\right) &= \mathbf{1}_N \otimes x_{avg}(t), \\
P_{Nn}(\mathscr{L} \otimes I_n) &= (\mathscr{L} \otimes I_n)P_{Nn}, \\
P_{Nn}(I_N \otimes A) &= (I_N \otimes A)P_{Nn},
\end{aligned}$$

we obtain the dynamics of $\mathbf{1}_N \otimes \hat{x}_{avg}(t)$ as follows:

$$\begin{aligned}
&\mathbf{1}_N \otimes \hat{x}_{avg}(t) \\
&= (I_{Nn} - \alpha(\mathscr{L} \otimes I_n))^L \left[(I_N \otimes A)(\mathbf{1}_N \otimes \hat{x}_{avg}(t-1)) + P_{Nn}\bar{y}(t)\right],
\end{aligned} \tag{1.31}$$

where $\bar{y}(t)$ is given in (1.29). By (1.31) and Lemma 1.2, the dynamics of $\bar{E}(t)$ is provided in the following:

$$\begin{aligned}\bar{E}(t+1) &= (I_{Nn} - \alpha(\mathscr{L} \otimes I_n) - P_{Nn})^L (I_N \otimes A)\bar{E}(t) \\ &\quad + (I_{Nn} - \alpha(\mathscr{L} \otimes I_n))^L (I_{Nn} - P_{Nn})\bar{y}(t) \\ &= (I_N \otimes A)(I_{Nn} - \alpha(\mathscr{L} \otimes I_n) - P_{Nn})^L \bar{E}(t) \\ &\quad + (I_{Nn} - P_{Nn})(I_{Nn} - \alpha(\mathscr{L} \otimes I_n))^L \bar{y}(t).\end{aligned} \tag{1.32}$$

According to Lemma 1.3, we have $\gamma < 1$. From Lemma 1.4, taking the norm on both sides of (1.32) yields

$$\left\|\bar{E}(t+1)\right\| \leq \|A\| \gamma^L \left\|\bar{E}(t)\right\| + \sqrt{N}\gamma^L \beta. \tag{1.33}$$

Under Assumption 1.1, $\left\|\bar{E}(t)\right\| = 0$. For a given L such that $L > \frac{\ln\|A\|}{\ln \gamma^{-1}}$, we have $\|A\| \gamma^L < 1$. Then recursively applying (1.33) for t times yields

$$\left\|\bar{E}(t)\right\| \leq (\|A\| \gamma^L)^t \left\|\bar{E}(0)\right\| + \sqrt{N}\gamma^L \beta \sum_{i=0}^{t-1} (\|A\| \gamma^L)^i \leq g(t). \tag{1.34}$$

The conclusion in (1.30) holds, due to $\|\bar{e}_i(t)\| \leq \sqrt{\sum_{j=1}^{N} \left\|\bar{e}_j(t)\right\|^2} = \left\|\bar{E}(t)\right\|$. □

Next, we study the boundedness of $\tilde{e}(t)$.

Lemma 1.6 *Consider Algorithm 2 satisfying Assumptions 1.1–1.4. By choosing* $\alpha = \frac{2}{\lambda_2(\mathscr{L})+\lambda_{max}(\mathscr{L})}$, *and* $L > \frac{\ln\|A\|}{\ln \gamma^{-1}}$, *the estimation error is asymptotically upper-bounded, i.e.,*

$$\limsup_{t\to\infty} \|\tilde{e}(t)\| \leq \frac{h(\beta, s)}{1 - \mu_0}, \tag{1.35}$$

if

$$h(\beta, s) < (1 - \mu_0) q_0, \tag{1.36}$$

where

$$\begin{aligned} h(\beta, s) &= \frac{s}{N}\beta + \frac{N-s}{N}(b_w + b_v + \|A\| g_0) + b_w, \\ g_0 &= \frac{\sqrt{N}\beta\gamma^L}{1 - \|A\| \gamma^L}, \\ \mu_0 &= \|A\| \left(1 - \frac{k^*}{N}\lambda_0\right), \\ k^* &= \min\left\{1, \frac{\beta}{\|A\| (g_0 + q_0) + b_w + b_v}\right\}. \end{aligned} \tag{1.37}$$

Proof By (1.31), the dynamics of the average estimate $\hat{x}_{avg}(t)$ are

$$\hat{x}_{avg}(t) = A\hat{x}_{avg}(t-1) + \frac{1}{N}(\mathbf{1}_N^\top \otimes I_n)P_{Nn}\,(I_{Nn} - \alpha(\mathscr{L} \otimes I_n))^L\,\bar{y}(t).$$

Due to

$$(\mathbf{1}_N^\top \otimes I_n)P_{Nn} = (\mathbf{1}_N^\top \otimes I_n)\,(I_{Nn} - \alpha(\mathscr{L} \otimes I_n))^L = (\mathbf{1}_N^\top \otimes I_n),$$

the average estimation error $\tilde{e}(t) = \hat{x}_{avg}(t) - x(t)$ satisfies

$$\begin{aligned}\tilde{e}(t) &= A\tilde{e}(t-1) - w(t-1)\\ &\quad + \frac{1}{N}(\mathbf{1}_N^\top \otimes I_n) \times \bar{C}^\top K_{\mathscr{I}}(t)(y(t) - \bar{C}(I_N \otimes A)\hat{X}(t-1))\\ &\quad + \frac{1}{N}(\mathbf{1}_N^\top \otimes I_n) \times \bar{C}^\top K_{\mathscr{I}^c}(t)(y(t) - \bar{C}(I_N \otimes A)\hat{X}(t-1))\\ &= M_t\tilde{e}(t-1) - \bar{m}_t + \tilde{m}_t,\end{aligned} \tag{1.38}$$

where the notations $K_{\mathscr{I}}(t)$ and $K_{\mathscr{I}^c}(t)$ are introduced in (1.13), $\mathscr{I}$ is a set of $N-s$ attack-free sensors, and

$$M_t = \left(I_n - \frac{1}{N}\sum_{i\in\mathscr{I}} k_i(t)C_i^\top C_i\right)A,$$

$$\bar{m}_t = w(t-1) + \frac{1}{N}(\mathbf{1}_N^\top \otimes I_n)\bar{C}^\top K_{\mathscr{I}}(t)\left(\bar{C}((I_N \otimes A)\bar{E}(t-1) - I_N \otimes w(t-1)) - v(t)\right),$$

$$\tilde{m}_t = \frac{1}{N}(\mathbf{1}_N^\top \otimes I_n)\bar{C}^\top K_{\mathscr{I}^c}(t)(y(t) - \bar{C}(I_N \otimes A)\hat{X}(t-1)).$$

For $\tilde{m}_t$, from (1.23) and Assumption 1.2, we have

$$\|\tilde{m}_t\| = \frac{1}{N}\left\|\sum_{i\in\mathscr{I}^c} C_i^\top k_i(t)(y_i(t) - C_iA\hat{x}_i(t-1))\right\| \le \frac{s}{N}\beta. \tag{1.39}$$

Regarding $\bar{m}_t$, by Assumption 1.2 and $k_i(t) \le 1$, it follows that

$$\begin{aligned}\|\bar{m}_t\| &\le \|w(t-1)\| + \frac{|\mathscr{I}|}{N}(\|A\|\,\|\bar{E}(t-1)\|\\ &\quad + \|I_N \otimes w(t-1)\| + b_v)\\ &\le \frac{N-s}{N}(b_w + b_v + \|A\|\,g_0) + b_w,\end{aligned} \tag{1.40}$$

where the last inequality is obtained by (1.34), and $g_0 = \sup_{t\ge 0} g(t)$.

Next, if (1.36) holds, we will prove the error $\tilde{e}(t)$ is uniformly bounded with $\|\tilde{e}(t)\| < q_0, \forall t \geq 1$. We use an induction method. By Assumption 1.1, at the initial time we have

$$\|\tilde{e}(0)\| = \left\|\hat{x}_{avg}(0) - x(0)\right\| \leq \frac{1}{N}\sum_{i=1}^{N}\left\|\hat{x}_i(0) - x(0)\right\| \leq q_0.$$

At time $t = 1$, for each attack-free sensor $i \in \mathscr{I}$,

$$\begin{aligned}|y_i(1) - C_i A\hat{x}_i(0)| &= | - C_i Ae_i(0) + C_i w(0) + v_i(1)|\\ &\leq \|A\| \, \|e_i(0)\| + b_w + b_v\\ &\leq \|A\| \left(\|\bar{e}_i(0)\| + \|\tilde{e}(0)\|\right) + b_w + b_v\\ &\leq \|A\| \left(g_0 + q_0\right) + b_w + b_v.\end{aligned} \tag{1.41}$$

Recall the form of $k_i(t)$, $i \in \mathscr{I}$, then we have

$$k_i(1) \geq k^* := \min\left\{1, \frac{\beta}{\|A\| \left(g_0 + q_0\right) + b_w + b_v}\right\} > 0.$$

Then

$$\begin{aligned}\|M_1\| &\leq \|A\| \left\|\left(I_n - \frac{k^*}{N}\sum_{i\in\mathscr{I}} C_i^{\mathrm{T}} C_i\right)\right\|\\ &\leq \|A\| \left(1 - \frac{k^*}{N}\lambda_{min}\left(\sum_{i\in\mathscr{I}} C_i^{\mathrm{T}} C_i\right)\right)\\ &= \|A\| \left(1 - \frac{k^*}{N}\lambda_0\right).\end{aligned} \tag{1.42}$$

By (1.38)–(1.40), we have $\|\tilde{e}(1)\| < q_0$.

Suppose the error $\tilde{e}(t-1)$ satisfies $\|\tilde{e}(t-1)\| < q_0$, for $t \geq 2$. At time t, for each attack-free sensor $i \in \mathscr{I}$, by repeating the proof process for $t = 1$ and using Lemma 1.5, we have $\|\tilde{e}(t)\| < q_0, t \geq 1$.

It follows from (1.38) that

$$\begin{aligned}\|\tilde{e}(t)\| &\leq \mu_0 \|\tilde{e}(t-1)\| + h(\beta, s)\\ &\leq \mu_0^t \|\tilde{e}(0)\| + h(\beta, s)\sum_{i=0}^{t-1} \mu_0^i,\end{aligned}$$

where μ_0 and $h(\beta, s)$ are provided in (1.37). Due to (1.36), $\mu_0 \in [0, 1)$, we have

$$\limsup_{t\to\infty} \|\tilde{e}(t)\| \le \frac{h(\beta, s)}{1-\mu_0}. \qquad \square$$

From Lemmas 1.5 and 1.6, the following theorem provides a condition such that the estimation error of DSF is asymptotically bounded.

Theorem 1.2 *Consider the system* (1.4) *and Algorithm 2. Under Assumptions 1.1–1.4, let* $\alpha = \frac{2}{\lambda_2(\mathscr{L})+\lambda_{max}(\mathscr{L})}$ *and* $L > \frac{\ln\|A\|}{\ln\gamma^{-1}}$, *then if*

$$h(\beta, s) < (1-\mu_0)q_0, \tag{1.43}$$

the estimation error of each sensor is asymptotically bounded, i.e., $\forall i \in \mathscr{V}$,

$$\limsup_{t\to\infty} \|e_i(t)\| \le \frac{h(\beta, s)}{1-\mu_0} + \frac{\sqrt{N}\beta\gamma^L}{1-\|A\|\gamma^L} < \infty, \tag{1.44}$$

where $h(\beta, s)$ *and* μ_0 *are given in* (1.37).

The condition (1.43) for the distributed filter corresponds to the condition (1.19) for the centralized filter. Compared to (1.19), (1.43) has one more term regarding the consensus error.

1.6 Numerical Simulations

In this section, we show the effectiveness of Algorithms 1 and 2 through numerical simulations.

The system state in (1.4) is observed by a sensor network with $N = 6$ nodes, where the communication topologies for the centralized and distributed frameworks are given in Figs. 1.3 and 1.4, respectively. Moreover, we assume

$$A = \begin{bmatrix} 1.01 & 0.1 \\ 0.1 & 1.05 \end{bmatrix}, \quad C_i = \begin{cases} \begin{bmatrix} 1 & 0 \end{bmatrix}, & \text{if } i \in \{1, 2, 3\}, \\ \begin{bmatrix} 0 & 1 \end{bmatrix}, & \text{if } i \in \{4, 5, 6\}. \end{cases}$$

Both the measurement noise $v_i(t), i = 1, \ldots, N$, and each element of the process noise $w(t)$ follow the uniform distribution in the range $[0, 1]$. The initial state and its estimate are $[1, 0]^\top$ and $[0, 0]^\top$, respectively. The simulation is implemented during the time interval $t = [0, 100]$ with time step 1. We simulate the algorithms by a Monte Carlo experiment with 100 runs. Suppose the adversary will insert the signal $a_i(t) = 3(C_i x(t) + v_i(t))$ to the measurement if sensor i is under attack. For Algorithm 1, denote the estimation error by

$$\eta(t) = \frac{1}{100}\sum_{j=1}^{100} \left\| e^j(t) \right\|,$$

where $e^j(t)$ stands for the estimation error by data center at time t in the jth run. For Algorithm 2, denote the estimation error of each sensor and the network respectively by

$$\eta_i(t) = \frac{1}{100}\sum_{j=1}^{100} \left\| e_i^j(t) \right\|, \quad \eta(t) = \frac{1}{100}\sum_{j=1}^{100} \max_{i\in\{1,\ldots,6\}} \left\| e_i^j(t) \right\|,$$

where $e_i^j(t)$ is the state estimation error by sensor i at time t in the jth run.

Given the consensus step $L = 1$ and the parameter $\beta = 3$, the system dynamics and the estimation errors of Algorithms 1 and 2 are shown in Fig. 1.5. As seen, each element of the system state is divergent and the estimation errors of Algorithms 1 and 2 are bounded. Moreover, the estimation error of the centralized filter, i.e., Algorithm 1, is smaller than that of the distributed filter in most sensors, i.e., Algorithm 2. Let $\beta = 3$, then the influence of consensus step L to the estimation error of Algorithm 2 is shown in Fig. 1.6, where a bigger consensus step L leads to a smaller estimation error. In addition, Algorithm 1 outweighs Algorithm 2 in estimation precision. In comparison with state values in Fig. 1.5, the values of the stable estimation errors are very small, which means the state is well estimated. Under $L = 2$, the relationships between the parameter β and the estimation error $\eta(t)$ for Algorithms 1 and 2 are studied in Figs. 1.7 and 1.8, respectively. Both Figs. 1.7 and 1.8 show that the estimation error of $\beta = 4$ is smaller than the errors of $\beta = 0.1$ and $\beta = 100$,

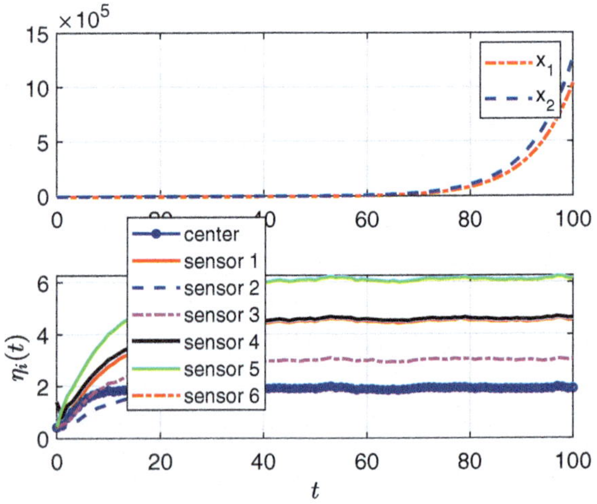

Fig. 1.5 The average system dynamics in the Monte Carlo experiment are shown in the above subfigure, where x_1 and x_2 stand for the first and second elements of the dynamics, respectively. In the below subfigure, the estimation errors of Algorithm 1 implemented by a data center, and Algorithm 2 by each sensor are shown

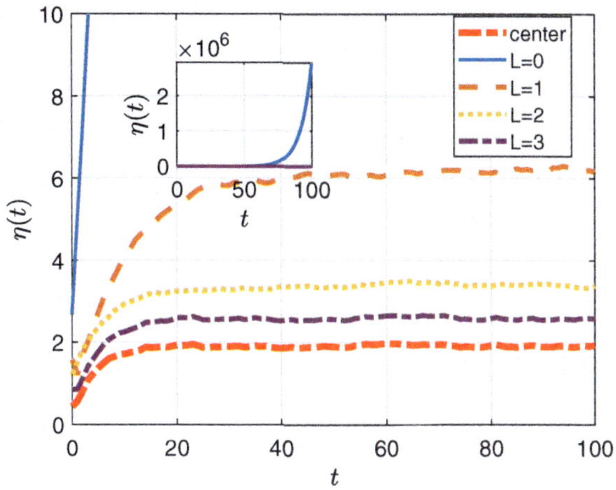

Fig. 1.6 The relationship between the estimation error $\eta(t)$ and the consensus step L of Algorithm 2 is shown. The estimation error of Algorithm 1 implemented by a data center is also provided for comparison

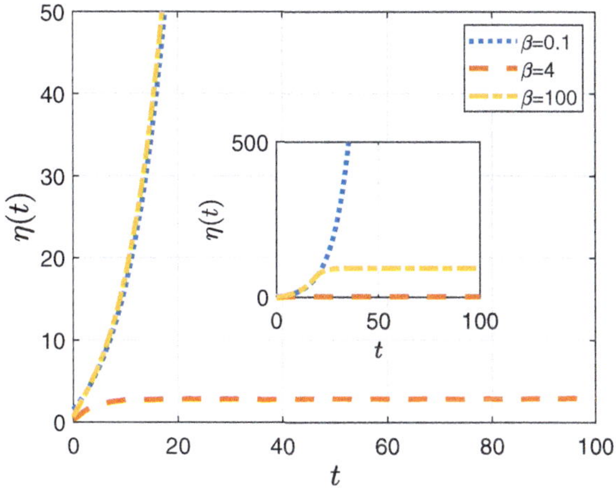

Fig. 1.7 The influence of β on the estimation error of Algorithm 1 is shown

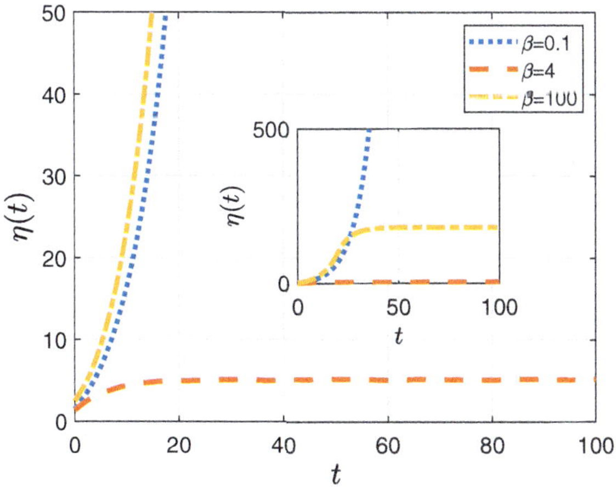

Fig. 1.8 The influence of β on the estimation error of Algorithm 2 is shown

which conforms to the previous discussion that a too small or too large β would lead to performance degradation. Comparing Figs. 1.7 and 1.8, when $\beta = 4$, Algorithm 1 still outweighs Algorithm 2 in estimation precision.

1.7 Conclusion and Future Work

This chapter studied the secure filtering problem for linear time-invariant systems with bounded noise under false data injection attacks in sensor networks, where an adversary can compromise a subset of sensors and manipulate the measurements arbitrarily. Two motivating examples from smart buildings and autonomous vehicles were given for this problem setup. Then by saturating the measurement innovations, we designed a secure centralized filter which is implemented by a data center with the measurements of all sensors. The method was then extended to a distributed version by employing a consensus operation in the state estimates of neighboring sensors. The estimation error of the distributed filter was shown to be asymptotically bounded. The developed filters can be utilized to improve the resilience and robustness of cyber-physical systems under potential sensor attacks.

There are some future directions of this work. One is to extend the saturation-based filter from LTI systems to more general systems, e.g., nonlinear systems. Another direction is to analyze the performance of the proposed filters under complex network or communication conditions, such as time-varying networks, time-delayed transmissions, and package missing. Last but not least, it is interesting to apply the saturation method to improve the resilience and robustness of controllers.

References

1. Chen, Y., Kar, S., Moura, J.M.: Resilient distributed field estimation. SIAM J. Control Optim. **58**(3), 1429–1456 (2020)
2. Deghat, M., Ugrinovskii, V., Shames, I., Langbort, C.: Detection and mitigation of biasing attacks on distributed estimation networks. Automatica **99**, 369–381 (2019)
3. Fawzi, H., Tabuada, P., Diggavi, S.: Secure estimation and control for cyber-physical systems under adversarial attacks. IEEE Trans. Autom. Control **59**(6), 1454–1467 (2014)
4. Forti, N., Battistelli, G., Chisci, L., Li, S., Wang, B., Sinopoli, B.: Distributed joint attack detection and secure state estimation. IEEE Trans. Signal Inf. Process. Netw. **4**(1), 96–110 (2018)
5. Han, D., Mo, Y., Xie, L.: Convex optimization based state estimation against sparse integrity attacks. IEEE Trans. Autom. Control **64**(6), 2383–2395 (2019)
6. He, X., Hashemi, E., Johansson, K.H.: Secure platooning of autonomous vehicles under attacked GPS data (2020). arXiv:2003.12975
7. He, X., Ren, X., Sandberg, H., Johansson, K.H.: Secure distributed filtering for unstable dynamics under compromised observations. In: Proceedings of IEEE Conference on Decision and Control, pp. 5344–5349 (2019)
8. Humayed, A., Lin, J., Li, F., Luo, B.: Cyber-physical systems security-a survey. IEEE Internet Things J. **4**(6), 1802–1831 (2017)

9. Kar, S., Moura, J.M., Poor, H.V.: Distributed linear parameter estimation: asymptotically efficient adaptive strategies. SIAM J. Control Optim. **51**(3), 2200–2229 (2013)
10. Lee, E.A.: Cyber physical systems: design challenges. In: Proceedings of International Symposium on Object and Component-Oriented Real-Time Distributed Computing, pp. 363–369 (2008)
11. Lee, J.G., Kim, J., Shim, H.: Fully distributed resilient state estimation based on distributed median solver. IEEE Trans. Autom. Control **65**(9), 3935–3942 (2020)
12. Li, K., Zhang, K.: Building occupancy estimation with robust Kalman filter. In: Proceedings of Asian Control Conference, pp. 1406–1410 (2017)
13. Mesbahi, M., Egerstedt, M.: Graph Theoretic Methods in Multiagent Networks. Princeton University Press (2010)
14. Mitra, A., Sundaram, S.: Byzantine-resilient distributed observers for LTI systems. Automatica **108**, 108487 (2019)
15. Nakahira, Y., Mo, Y.: Attack-resilient $\mathscr{H}_2$, $\mathscr{H}_\infty$, and $\mathscr{H}_1$ state estimator. IEEE Trans. Autom. Control **63**(12), 4353–4360 (2018)
16. Pajic, M., Lee, I., Pappas, G.J.: Attack-resilient state estimation for noisy dynamical systems. IEEE Trans. Control Netw. Syst. **4**(1), 82–92 (2017)
17. Pajic, M., Weimer, J., Bezzo, N., Sokolsky, O., Pappas, G.J., Lee, I.: Design and implementation of attack-resilient cyberphysical systems: with a focus on attack-resilient state estimators. IEEE Control Syst. Mag. **37**(2), 66–81 (2017)
18. Pasqualetti, F., Dörfler, F., Bullo, F.: Attack detection and identification in cyber-physical systems. IEEE Trans. Autom. Control **58**(11), 2715–2729 (2013)
19. Ren, X., Mo, Y., Chen, J., Johansson, K.H.: Secure state estimation with Byzantine sensors: a probabilistic approach. IEEE Trans. Autom. Control **65**(9), 3742–3757 (2020)
20. Shoukry, Y., Chong, M., Wakaiki, M., Nuzzo, P., Sangiovanni-Vincentelli, A., Seshia, S.A., Hespanha, J.P., Tabuada, P.: SMT-based observer design for cyber-physical systems under sensor attacks. ACM Trans. Cyber-Phys. Syst. **2**(1), 5 (2018)
21. Shoukry, Y., Nuzzo, P., Puggelli, A., Sangiovanni-Vincentelli, A.L., Seshia, S.A., Tabuada, P.: Secure state estimation for cyber-physical systems under sensor attacks: a satisfiability modulo theory approach. IEEE Trans. Autom. Control **62**(10), 4917–4932 (2017)

Chapter 2
Zero-Dynamics Attack, Variations, and Countermeasures

Hyungbo Shim, Juhoon Back, Yongsoon Eun, Gyunghoon Park, and Jihan Kim

Abstract This chapter presents an overview of actuator attacks that exploit zero dynamics, and countermeasures against them. First, zero-dynamics attack is reintroduced based on a canonical representation called normal form. Then it is shown that the target dynamic system is at elevated risk if the associated zero dynamics is unstable. From there on, several questions are raised in series to ensure when the target system is immune to an attack of this kind. The first question is: *Is the target system secure from zero-dynamics attack if it does not have any unstable zeros?* An answer provided for this question is: *No, the target system may still be at risk due to another attack surface emerging in the process of implementation.* This is followed by a series of questions, and in the course of providing answers, variants of the classic zero-dynamics attack are presented, from which the vulnerability of the target system is explored in depth. In the end, countermeasures are proposed to render the attack ineffective. Because it is known that zero dynamics in continuous-time systems cannot be modified by feedback, the main idea of the countermeasure is to relocate any unstable zero to a stable region in the stage of digital implementation through modified digital samplers and holders. Adversaries can still attack actuators, but due to the relocated zeros, they are of little use in damaging the target system.

H. Shim (✉) · J. Kim
Seoul National University, Seoul, Korea
e-mail: hshim@snu.ac.kr

J. Kim
e-mail: jihan89@cdsl.kr

J. Back
Kwangwoon University, Seoul, Korea
e-mail: backhoon@kw.ac.kr

Y. Eun
DGIST, Daegu, Korea
e-mail: yeun@dgist.ac.kr

G. Park
KIST, Seoul, Korea

H. Ishii and Q. Zhu (eds.), *Security and Resilience of Control Systems*,
Lecture Notes in Control and Information Sciences 489,
https://doi.org/10.1007/978-3-030-83236-0_2

2.1 Introduction

Modern control systems embrace the technological advances in communication and computing in order to increase their capability and applicability: feedback control systems now handle large-scale systems whose subsystems are not necessarily collocated, and play key roles in societal infrastructures such as electrical power distribution facilities, oil and gas pipelines, water distribution systems, and sewage treatment plants. The wide usage of Supervisory Control And Data Acquisition (SCADA) systems to monitor and control operations of the large-scale plant is an example. However, what also increased along with the capability and applicability is the *vulnerability* of the systems. Specifically, technological advances over the past decade have seen these traditionally closed systems become open and Internet-connected, which puts the system at risk from attacks with malicious intent. Incidents of Iranian nuclear facility [13], breach in Maroochy water sewage [24], and massive outage in Ukrainian power grid [14] show that the risk is not a possibility but a reality. Clearly, the effect of malfunction induced by malicious attack on control systems could be catastrophic as witnessed in [4, 13, 14, 21, 24]. Naturally, security of control systems has become an active topic of research due to its importance.

Figure 2.1 shows the modern control systems under consideration, where remotely located sensors and actuators are connected to computing units that execute feedback control algorithms, monitoring of the operation, and a level of anomaly detection. Here, values of sensor measurement, $y(t)$, are sent through the communication network and so are those for the actuators, $u(t)$. Attacks on sensors, denoted by $a_s(t)$, potentially corrupt sensor measurement $y(t)$ and so do the attacks on actuators, denoted by $a_a(t)$, the actuation signals for the plant.

The community of automatic control began to investigate not only countermeasures to malicious attacks but also mechanisms of various potential attacks on sensors, actuators, and controllers themselves [15, 16, 18–20]. What appears to be particularly dangerous is a class of attacks that are referred to be *stealthy*. An attack being stealthy means that its effect does not appear in the output of the system or the

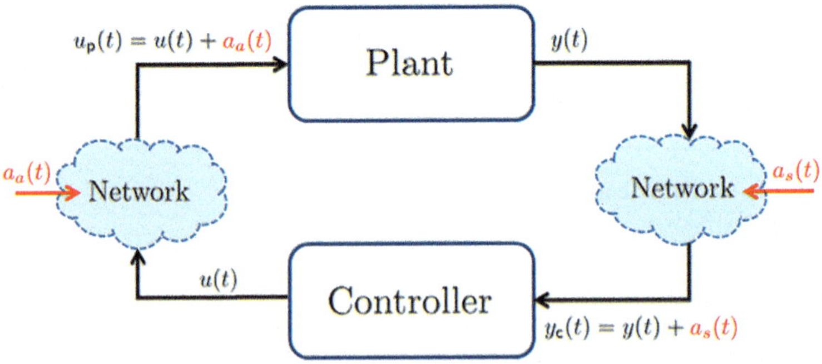

Fig. 2.1 Illustration of modern control systems with remote communication under a sensor attack and an actuator attack

output of the anomaly detector which is often deployed for the purpose of attack detection. The most well-known one in this class is the zero-dynamics attack [26]: it is an actuator attack that pushes the system states along the zero dynamics. Due to the intrinsic property of zero dynamics, no sign shows in the output, which makes this attack stealthy. Another notion that we would like to bring up is an attack being *disruptive*. This means that a portion of the system states diverges due to the attack. A zero-dynamics attack is disruptive if the associated zero dynamics is unstable.

The central topic of this chapter is on the attacks that exploit the zero dynamics and also their countermeasures. Specifically, it is shown that a zero-dynamics attack is feasible even without precise knowledge of the target systems (which is not the case in [26]). Also shown is that it may be too soon to conclude that the systems having no zero dynamics are unsusceptible to such attacks: Sampling may bring a new vulnerability, and unstable pole dynamics may pose a similar threat if sensor attacks (as opposed to actuator attacks) are considered. Finally, we discuss zero-dynamics attacks for nonlinear systems. As potential countermeasures, we introduce methods of moving zeros by generalized hold or by generalized sampling. Two methods move unstable zeros that may be targets for the adversaries, to a stable region in the complex plane. Then, although the systems can still be under stealthy attacks, the attacks lose effectiveness and the adversaries lose motivation. All the results described are illustrated with numerical examples to help readers grasp the concepts.

2.2 Threats of Zero-Dynamics Attack

Among many potential threats to the control systems, the zero-dynamics attack is known to be lethal because it is fundamentally difficult to detect the attack. Detection of the attack often takes place at the controller side, and so, the information available for detection is the input data transmitted from the controller and the output data received from the plant, both of which may be compromised over the communication networks. Based on this fact, the attacker deceives the controller in a way that the transmitted input (u in Fig. 2.1) and the received output ($y_{\mathsf{c}} = y + a_s$ in Fig. 2.1) would appear to be consistent with the dynamics of the plant.

From an attacker's perspective, attack specification consists of initiation time $t_0 > 0$, hazard level L_{hazard}, and detection level L_{detect}. With them, it is said that the attack is *disruptive* if there exists $t^* \geq t_0$ such that

$$\|x(t^*) - x_{\mathsf{af}}(t^*)\| \geq L_{\mathsf{hazard}} \tag{2.1}$$

where x is the actual plant's state and x_{af} is the attack-free state that would have resulted if there were no attacks. At the same time, the attack should be *stealthy*, which means that

$$\|y_{\mathsf{c}}(t) - y_{\mathsf{af}}(t)\| \leq L_{\mathsf{detect}} \quad \text{for all } t \text{ such that } t_0 \leq t \leq t^* \tag{2.2}$$

where y_{c} is the received output and y_{af} is the attack-free output that would have resulted if there were no attacks. Violation of (2.2) indicates that the system is under attack, and the level of detection L_{detect} is considered to be set by the defender, which is not to be zero for avoiding false alarm due to measurement noise and other perturbation. In practice, rather than y_{af} (which is not simple to compute), the estimate $\hat{y}$ of y is often used for attack detection where $\hat{y}$ is computed by the controller using the history of the input (sent to the plant) and the output (received from the plant) with the system model. The idea behind the replacement of y_{af} with $\hat{y}$ is the following: If the estimator (that generates the estimate $\hat{y}$) is continuous with respect to its input y_{c} (for example, a dynamic system such as Luenberger observer), then (a) $\hat{y}(t) = y_{\mathsf{af}}(t)$ if the input to the estimator is y_{af}; (b) $\hat{y}(t) \approx y_{\mathsf{af}}(t)$ if $y_{\mathsf{c}}(t) \approx y_{\mathsf{af}}(t)$ by continuity; and (c) $\|y_{\mathsf{c}}(t) - \hat{y}(t)\| \le \|y_{\mathsf{c}}(t) - y_{\mathsf{af}}(t)\| + \|y_{\mathsf{af}}(t) - \hat{y}(t)\|$, so that an attack that yields the compromised output $y_{\mathsf{c}}(t)$ arbitrarily close to $y_{\mathsf{af}}(t)$ can make $\|y_{\mathsf{c}}(t) - \hat{y}(t)\|$ arbitrarily small and thus is stealthy.

In this section, the goal of the attacker is to design an attack signal that is both disruptive and stealthy. We restrict the focus on attacks that exploit zero dynamics of the target systems.

2.2.1 *Zero-Dynamics Attack*

The introduction of the zero-dynamics attack dates back at least to [26], where a geometric notion is employed to illustrate the idea of the zero-dynamics attack. In this subsection, the zero-dynamics attack is re-interpreted in view of a canonical form that is a special realization of an LTI system.

Consider a continuous-time SISO LTI system represented by the following transfer function of relative degree r:

$$G(s) = \frac{\beta_{n-r}s^{n-r} + \cdots + \beta_1 s + \beta_0}{s^n + \alpha_{n-1}s^{n-1} + \cdots + \alpha_0}.$$

It is known (from, e.g., [8, Example 13.4]) that a state-space minimal realization of $G(s)$, called the (Byrnes–Isidori) normal form, is given by

$$\begin{aligned} y = x_1 \qquad \dot{x}_1 &= x_2 \\ \dot{x}_2 &= x_3 \\ &\vdots \\ \dot{x}_{r-1} &= x_r \\ \dot{x}_r &= \phi_:^\top x_: + \phi_z^\top x_z + bu \\ \dot{x}_z &= Sx_z + px_1, \end{aligned}$$

where y is the output, u is the input, and the state x consists of $x_{:} = [x_1, x_2, \dots, x_r]^\top \in \mathbb{R}^r$ and $x_z \in \mathbb{R}^{n-r}$. The system parameters are $\phi_{:} \in \mathbb{R}^r$, $\phi_z \in \mathbb{R}^{n-r}$, $S \in \mathbb{R}^{(n-r)\times(n-r)}$, $p \in \mathbb{R}^{(n-r)\times 1}$, and $b \in \mathbb{R}$. Here, it can be shown that the roots of zero polynomial $\beta_{n-r}s^{n-r} + \cdots + \beta_1 s + \beta_0$ coincide with the eigenvalues of the matrix $S \in \mathbb{R}^{(n-r)\times(n-r)}$ (see [8, Example 13.4]).

About zero dynamics: Zero dynamics of a system is the residual subdynamics that remains when the output maintains identically zero by a pair of some initial condition and some input. A benefit of the normal form is that the zero dynamics is easily identified, which is simply $\dot{x}_z = Sx_z$. The pair of initial condition and input that makes the output identically zero is also identified as $(x_{:}(0), x_z(0)) = (0, \Delta)$ and $u(t) = -(1/b)\phi_z^\top x_z(t)$ where Δ is arbitrary. Rigorously speaking, the dynamics $\dot{x}_z = Sx_z + px_1$ is called "internal dynamics", but is sometimes called "zero dynamics" by abuse.

Suppose that an LTI dynamic controller $C(s) = H(sI - F)^{-1}G$ internally stabilizes the closed-loop system, and that an adversary can inject malicious signal $a_a(t)$ (called "actuator attack") into the input channel. Then, the closed-loop system with the controller can be written as[1]

$$\begin{aligned} y = x_1 \qquad \dot{x}_{:} &= O^1 x_{:} + 0_1(\phi_{:}^\top x_{:} + \phi_z^\top x_z + b(Hc + a_a)) \\ \dot{x}_z &= Sx_z + px_1 \\ \dot{c} &= Fc + Gx_1 \end{aligned} \tag{2.3}$$

where c is the state of the controller. Under the stability assumption, the following inequality holds for the system in (2.3):

$$\left\| \begin{bmatrix} x_{:}(t) \\ x_z(t) \\ c(t) \end{bmatrix} \right\| \le ke^{-\lambda t} \left\| \begin{bmatrix} x_{:}(0) \\ x_z(0) \\ c(0) \end{bmatrix} \right\| \tag{2.4}$$

with some positives k and λ.

The following proposition is another interpretation of [26] in terms of the normal form.

[1] We use the notation O^1 and 0_1 (of suitable size) as

$$O^1 = \begin{bmatrix} 0 & 1 & 0 & \cdots & 0 \\ 0 & 0 & 1 & \cdots & 0 \\ \vdots & \vdots & \vdots & \ddots & \vdots \\ 0 & 0 & 0 & \cdots & 1 \\ 0 & 0 & 0 & \cdots & 0 \end{bmatrix} \quad \text{and} \quad 0_1 = \begin{bmatrix} 0 \\ 0 \\ \vdots \\ 0 \\ 1 \end{bmatrix}.$$

Proposition 2.1 *Suppose that an actuator attack signal a_a is generated by*

$$\dot{z}_a = S z_a, \qquad a_a = -\frac{1}{b}\phi_z^\top z_a \tag{2.5}$$

with an initial condition $z_a(t_0) = \Delta$, and is injected from the time $t_0 > 0$ into the closed loop. If $\|\Delta\|$ is sufficiently small, then the attack becomes stealthy. Moreover, if S has at least one eigenvalue whose real part is positive (i.e., the plant has an unstable zero, or, is non-minimum phase) and the initial condition Δ excites the unstable mode, then the attack is disruptive.

Proof While the attack generator is valid only for $t \geq t_0$, it can be equivalently rewritten as $\dot{z}_a = S z_a + \Delta\delta(t - t_0)$ with $z_a(0) = 0$, which is valid for $t \geq 0$, where δ is Dirac's delta function. Then, with $\tilde{x}_z := x_z - z_a$, the closed-loop system can be rewritten as

$$\begin{aligned} y = x_1 \qquad \dot{x}_{:} &= O^1 x_{:} + 0_1(\phi_{:}^\top x_{:} + \phi_z^\top \tilde{x}_z + bHc) \\ \dot{\tilde{x}}_z &= S\tilde{x}_z + p x_1 - \Delta\delta(t - t_0) \\ \dot{c} &= Fc + Gx_1. \end{aligned} \tag{2.6}$$

By comparing (2.3) (without a_a) and (2.6), it is seen that they are the same except the replacement of x_z with $\tilde{x}_z$ and the perturbation Δ. The perturbation causes a transient response after t_0. Because the attack-free response x_{af} in (2.2) is the same as the response of (2.6) with $\Delta = 0$, and because the system (2.6) is stable, it is seen that $y(t)$ (which is the same as $y_c(t)$ since there is no sensor attack) can be made arbitrarily close to $y_{af}(t)$ (which is the first component of $x_{af}(t)$) by choosing $\|\Delta\|$ small enough. Hence, the attack is stealthy (for any t^* in (2.2)). Now, by choosing a small Δ that excites the unstable mode of (2.5), the state $x_z(t)$ diverges because $z_a(t)$ diverges while $\tilde{x}_z(t) = x_z(t) - z_a(t)$ exponentially converges to zero. This shows the attack is disruptive. □

Remark 2.1 If S is expressed in a coordinate where $S = \text{blockdiag}(S_u, S_s)$ where all the eigenvalues of S_s have positive real parts, then the dimension of attack generator (2.5) can be reduced as $\dot{z}_a = S_s z_a$ and $a_a = -(1/b)\phi_{z,s}^\top z_a$ where $\phi_{z,s}$ is the corresponding part of ϕ_z.

2.2.1.1 Pole-Dynamics Attack

As a variation of the zero-dynamics attack, which is an actuator attack, we consider its dual "pole-dynamics attack" that is a sensor attack studied in [7]. If we rewrite the controller using the normal form, the closed-loop system is written as

$$u = c_1 \quad \begin{aligned} \dot{c}_: &= O^1 c_: + 0_1(\phi_:^\top c_: + \phi_z^\top c_z + b(y + a_s)) \\ \dot{c}_z &= S c_z + p c_1 \\ \dot{x} &= Ax + Bc_1, \qquad y = Cx, \end{aligned} \tag{2.7}$$

where x is the state of the plant $C(sI - A)^{-1}B$, and $\phi_:$, ϕ_z, b, S, and p are the parameters of the controller. By comparison, it is seen that the same argument holds for the design of the sensor attack a_s.

Proposition 2.2 *Suppose that a sensor attack signal a_s is generated by*

$$\dot{z}_\mathsf{a} = A z_\mathsf{a}, \qquad a_s = -C z_\mathsf{a} \tag{2.8}$$

with an initial condition $z_\mathsf{a}(t_0) = \Delta$, and is injected from the time $t_0 > 0$ into the closed loop. If $\|\Delta\|$ is sufficiently small, then the attack becomes stealthy. Moreover, if A has at least one eigenvalue whose real part is positive (i.e., the plant has an unstable pole, or the plant is unstable) and the initial condition Δ excites the unstable mode, then the attack is disruptive.

Proof With $\tilde{x} := x - z_\mathsf{a}$, we have

$$u = c_1 \quad \begin{aligned} \dot{c}_: &= O^1 c_: + 0_1(\phi_:^\top c_: + \phi_z^\top c_z + bC\tilde{x}) \\ \dot{c}_z &= S c_z + p c_1 \\ \dot{\tilde{x}} &= A\tilde{x} + Bc_1 - \Delta\delta(t - t_0) \end{aligned} \tag{2.9}$$

so that $(x(t) - z_\mathsf{a}(t))$ exponentially converges to zero after t_0, so that $x(t)$ diverges if Δ excites unstable mode of A. Hence, the attack is disruptive. At the same time, with sufficiently small Δ, the states of (2.9) can be made arbitrarily close to the states of (2.7) without a_s. Therefore, the received output $y_\mathsf{c} = y - a_s = Cx - Cz_\mathsf{a} = C\tilde{x}$ can be made arbitrarily close to the output y of (2.7) without a_s which corresponds to y_af. Hence, the attack is stealthy by (2.2). □

The zero-dynamics attack is a stealthy attack regardless of whether the plant has unstable zeros or not. However, the attack becomes disruptive when there is an unstable zero. At this moment, a natural question arises: "Does it mean that the control system is secure from disruptive stealthy attacks if the plant has no unstable zeros?"

2.2.2 *Sampling-Zero-Dynamics Attack*

Unfortunately, a disruptive and stealthy attack that exploits zero dynamics can be designed even if there are no unstable zeros *in continuous time*. Modern control systems are mostly implemented by a digital controller, and thus, the continuous-time output signal of the plant is sampled to a discrete-time signal, and the discrete-

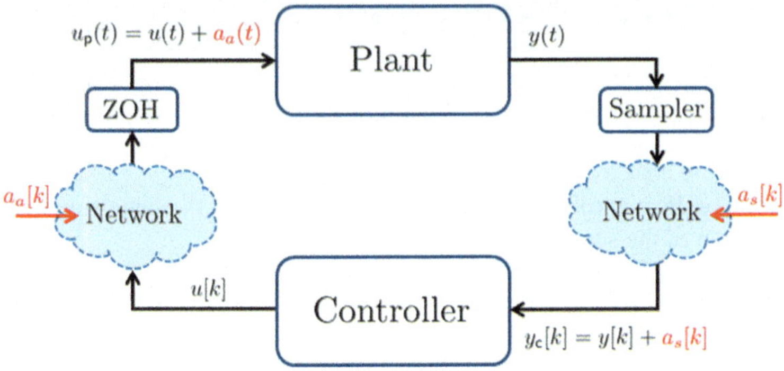

Fig. 2.2 Control system implemented with zero-order hold and sampler

time control input is converted to a continuous-time signal by zero-order hold (see Fig. 2.2). The role of the sampler and zero-order hold is defined as

- sampler: $y[k] := y(kT_s)$ and
- zero-order hold: $u(t) = u[k]$, $kT_s \le t < (k+1)T_s$,

where T_s is the sampling period and k is the discrete-time index.

Suppose that the continuous-time plant is given by $G(s)$ whose relative degree is r. Then, the representation of the plant from the discrete-time input $u[k]$ to output $y[k]$ becomes the discrete-time transfer function $\bar{G}(z)$. It is emphasized that the relative degree $\bar{r}$ of $\bar{G}(z)$ is equal to 1 for almost all sampling periods T_s, regardless of the relative degree r of $G(s)$. This means that sampling introduces $r - 1$ additional zeros, and these new zeros are called *sampling zeros*. The bad news is that if $r \ge 3$ and the sampling is fast so that T_s is small, then at least one of the sampling zeros is always unstable, and so, the attacker can deploy the *sampling-zero-dynamics attack* even if the continuous-time plant has no unstable zeros.

About sampling zeros: Let us first ask why the discrete-time plant $\bar{G}(z)$ has $r - 1$ additional zeros over its continuous-time counterpart $G(s)$ for almost all T_s. To see this, let us examine a typical step response of, e.g., $G(s) = (-s+3)/(s^3+3s^2+3s+1)$ as follows:

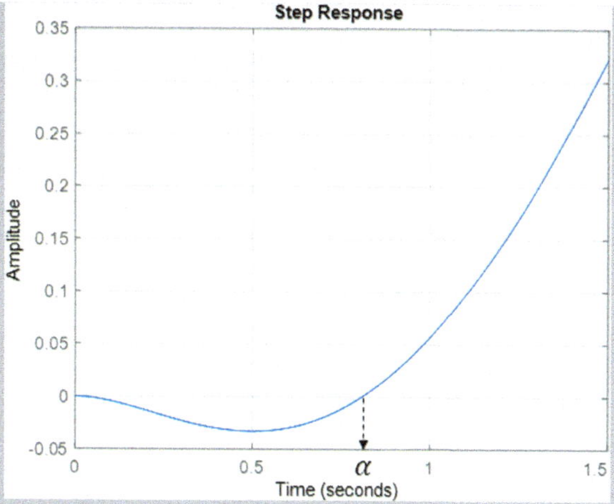

Since the discrete-time step input $u[k]$ is converted as a unit-step continuous-time input $u(t)$, the discrete-time step response $y[k]$ can be found from the plot, that is, $y[k] = y(kT_s)$. Then, from the plot, it is seen that $y[1] = y(T_s) \neq 0$ for almost all T_s, i.e., except $T_s = \alpha$ in the plot. This means that the relative degree of $\bar{G}(z)$ is 1 except the case $T_s = \alpha$, even if the relative degree of $G(s)$ is $r = 2$. Thus, there should be one sampling zero in $\bar{G}(z)$.

In fact, the following formal statement is taken from [27]:

Lemma 2.1 *For almost all $T_s > 0$,*

$$\bar{G}(z) = g\frac{\gamma_{r-1}(z)\prod_{i=1}^{n-r}(z-\bar{z}_i)}{\prod_{i=1}^{n}(z-\bar{p}_i)},$$

where

- $\bar{p}_i = e^{\lambda_i T_s}$ *is the pole where* λ_i *is the pole of* $G(s)$.
- $\bar{z}_i$ *is the zero (When* T_s *is sufficiently small,* $\bar{z}_i$ *gets close to* $e^{\mu_i T_s}$ *where* μ_i *is the zero of* $G(s)$*. Since these zeros appear from the continuous-time zeros, they are called "intrinsic zeros".).*
- $\gamma_{r-1}(z)$ *is a polynomial of order* $r-1$*, dependent on the sampling time* T_s *(As* $T_s \to 0$*, the coefficients of* γ_{r-1} *converge to those of the Euler–Frobenius polynomial* $P_{r-1}(z)$ *of order* $r-1$*. The roots of* $\gamma_{r-1}(z)$ *are called "sampling zeros".).*
- g *is a constant defined by* $(T_s)^r(\lim_{s\to\infty} s^r G(s))/r!$*.*

The Euler–Frobenius polynomial of order $r-1$ is defined as $P_{r-1}(z) = \beta_{r-1}z^{r-1} + \beta_{r-2}z^{r-2} + \cdots + \beta_1 z + \beta_0$ where $\beta_i = \sum_{j=0}^{i}(-1)^{i-j}(j+1)^r\binom{r+1}{i-j}$, which is known to have at least one unstable root that is located outside of the unit circle in the complex plane whenever $r \geq 3$ [27]. Therefore, if the continuous-time plant $G(s)$ has a relative degree $r \geq 3$ and the sampling period T_s is sufficiently small, then the appearance of unstable sampling zero is unavoidable. All the roots of the Euler–Frobenius polynomial are negative real. An implication is that the output-zeroing input corresponding to the sampling zero should have alternating signs.

Since the relative degree of $\bar{G}(z)$ is 1 for almost all cases, let us suppose that $\bar{G}(z)$ is realized in the (discrete-time) normal form:

$$y[k] = x_1[k] \qquad \begin{aligned} x_1[k+1] &= \phi_{:}x_1[k] + \phi_z^{\top} x_z[k] + b(u[k] + a_a[k]) \\ x_z[k+1] &= Sx_z[k] + px_1[k] \end{aligned} \tag{2.10}$$

where $x_z \in \mathbb{R}^{n-1}$.

Proposition 2.3 *Suppose that an actuator attack signal $a_a[k]$ is generated by*

$$z_a[k+1] = Sz_a[k], \qquad a_a[k] = -\frac{1}{b}\phi_z^{\top} z_a[k] \tag{2.11}$$

with $z_a[k_0] = \Delta$ where the attack initiates at $t_0 = k_0 T_s$. If $\|\Delta\|$ is sufficiently small, then the attack becomes stealthy. Moreover, if S has at least one eigenvalue outside of the unit circle (which is either an intrinsic zero or a sampling zero) and the initial condition Δ excites the unstable mode, then the attack is disruptive.

Proof Define $\tilde{x}_z := x - z_a$ and consider the controller is given by $\bar{C}(z) = H(zI - F)^{-1}G$. Then, the closed-loop system becomes

$$y[k] = x_1[k] \qquad \begin{aligned} x_1[k+1] &= \phi_{:}x_1[k] + \phi_z^{\top} \tilde{x}_z[k] + bHc[k] \\ \tilde{x}_z[k+1] &= S\tilde{x}_z[k] + px_1[k] - \Delta\delta[k - k_0] \\ c[k+1] &= Fc[k] + Gx_1[k] \end{aligned} \tag{2.12}$$

where δ is Kronecker's delta. The rest of the argument is similar to the proof of Proposition 2.1. □

Example 2.1 The transfer function of Fig. 2.3 from $u(t)$ to $y(t) = x_1(t)$ is $G(s) = (s+1)/(s^4 + 2s^3 + 3s^2 + s + 1)$, which is a minimum phase system having relative degree 3. With $T_s = 0.1$, its discrete-time equivalent model can be written as (2.10) with

$$S = \begin{bmatrix} 0 & 1 & 0 \\ 0 & 0 & 1 \\ 0.86 & 2.58 & -2.99 \end{bmatrix}, \quad p = \begin{bmatrix} 0 \\ 0 \\ 1 \end{bmatrix}, \quad \phi_z = \begin{bmatrix} 5.02 \\ 20.04 \\ -28.3 \end{bmatrix}$$

$$\phi_{:} = 6.78, \quad b = 1.62 \times 10^{-4}.$$

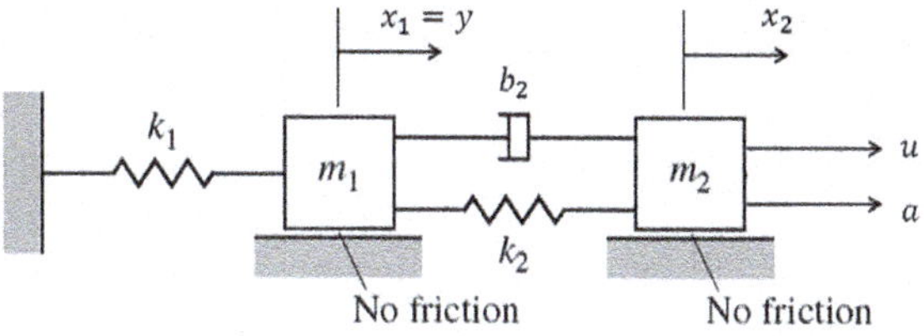

Fig. 2.3 Two-mass system with all parameters set to 1. (Reprinted from [11], Copyright 2020, with permission from Elsevier)

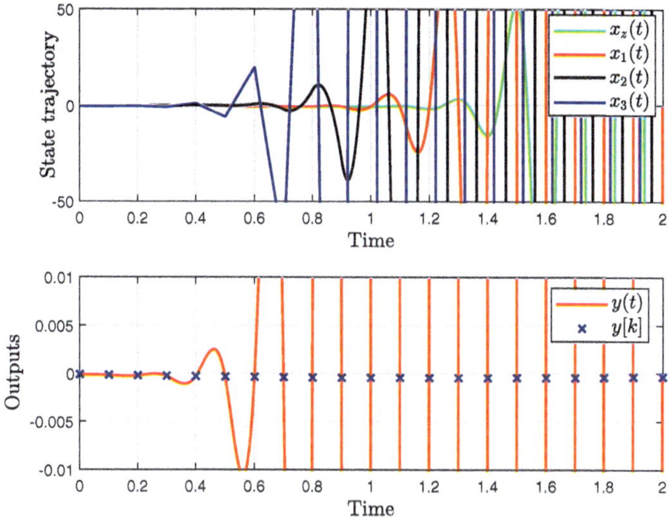

Fig. 2.4 The first plot is the state trajectory and the second is the outputs. Red color is the output $y(t)$, and the sampled output $y[k]$ is marked as blue cross. (Reprinted from [11], Copyright 2020, with permission from Elsevier)

Note that the eigenvalues of S are -3.64, -0.26, and 0.90, of which the first two are sampling zeros. Clearly, there is an unstable zero of -3.64. With $\Delta = 10^{-5} \times [1, 0, 0]^T$, the attack generated by (2.11) results in Fig. 2.4. □

Up to now, we have seen that even if the continuous-time plant does not have unstable zeros, sampling for discrete-time implementation of the control system may introduce unstable sampling zeros in a discrete-time domain. At this moment, a natural question arises: "If the discrete-time plant has no unstable (sampling) zero, does it mean that the control system is secure from disruptive stealthy attacks?"

2.2.3 *Enforced-Zero-Dynamics Attack*

Unfortunately, even if the plant has no unstable zero in the discrete-time domain, there is some possibility to design a disruptive stealthy attack when there is more freedom in actuation than sensing. Since sensing occurs only at discrete time instances, if an actuator attack drives the plant's state to pass through the kernel of the output matrix at each sensing time while the state is enforced to behave disruptively, then the attack can remain stealthy while the state diverges (see Fig. 2.5, for example). This type of attack exists when the sampled-data system has an input redundancy, i.e., the number of inputs being larger than that of outputs and/or the sampling rate of the actuators being higher than that of the sensors. For the sake of brevity, only the latter case is explained in this chapter in detail.

Consider a linear time-invariant plant

$$\begin{aligned} \dot{x}(t) &= Ax(t) + B(u(t) + a_a(t)) \quad \in \mathbb{R}^n, \qquad u(t),\ a_a(t) \quad \in \mathbb{R}^p, \\ y(t) &= Cx(t) \quad \in \mathbb{R}^q, \end{aligned} \tag{2.13}$$

where a_a is the actuator attack, and the matrices B and C are assumed to have full column rank and full row rank, respectively. Let T_s be the sampling period for the output measurement such that

$$y[k] = y(kT_s)$$

and T_a be the period for zero-order hold such that

$$u(t) + a_a(t) = u[i] + a_a[i], \quad iT_a \le t < (i+1)T_a$$

in which, it is assumed that a discrete-time attack $a_a[i]$ is injected into the communication network so that it is converted together with $u[i]$. Then, the enforced-zero-dynamics attack $a_a[i]$ can be generated under the assumption that

$$qT_a < pT_s. \tag{2.14}$$

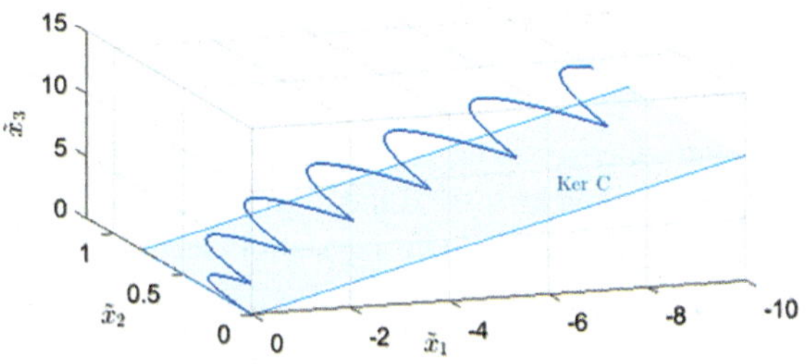

Fig. 2.5 Solution diverges from the origin while it belongs to the kernel of the output matrix C at each sampling instance

It should be noted that (2.14) holds if the number of inputs, p, is large and/or the sampling period for actuation, T_a, is small. Based on (2.14), a disruptive stealthy attack is proposed in [12], and the attack is termed as "masking attack".

In order to introduce the idea of the masking attack (we call it "enforced-zero-dynamics attack" in this chapter), let us simplify the situation as

$$p = q = 1, \qquad T_s = 2T_a \tag{2.15}$$

and refer to [12] for the general cases. Now, let

$$A_{\mathsf{d}} := e^{AT_a} \quad \text{and} \quad B_{\mathsf{d}} := \int_0^{T_a} e^{A\tau} d\tau \; B.$$

Also, let $\tilde{x} := x - x_{\mathsf{af}}$ and $\tilde{y} := y - Cx_{\mathsf{af}}$ where x_{af} is the solution of (2.13) when there is no attack, and let the attack initiate at $t_0 = 0$ (so that $\tilde{x}(0) = 0$). Then, it is seen that

$$\begin{bmatrix} \tilde{x}(T_a) \\ \tilde{x}(2T_a) \end{bmatrix} = \begin{bmatrix} B_{\mathsf{d}} & 0 \\ A_{\mathsf{d}} B_{\mathsf{d}} & B_{\mathsf{d}} \end{bmatrix} \begin{bmatrix} a_a[0] \\ a_a[1] \end{bmatrix} \tag{2.16a}$$

$$\tilde{y}[1] = \tilde{y}(T_s) = \tilde{y}(2T_a) = \begin{bmatrix} CA_{\mathsf{d}}B_{\mathsf{d}} & CB_{\mathsf{d}} \end{bmatrix} \begin{bmatrix} a_a[0] \\ a_a[1] \end{bmatrix}. \tag{2.16b}$$

Now, we assume that

$$\text{(a) } \ker \begin{bmatrix} CA_{\mathsf{d}}B_{\mathsf{d}}, & CB_{\mathsf{d}} \end{bmatrix} \not\subset \ker \begin{bmatrix} B_{\mathsf{d}} & 0 \\ A_{\mathsf{d}} B_{\mathsf{d}} & B_{\mathsf{d}} \end{bmatrix},$$
$$\text{(b) } \operatorname{im} CA_{\mathsf{d}}^2 \subset \operatorname{im} \begin{bmatrix} CA_{\mathsf{d}}B_{\mathsf{d}}, & CB_{\mathsf{d}} \end{bmatrix},$$

which automatically follow for almost all sampling periods from the assumption (2.15) and that the matrix B has full column rank. Indeed, item (a) holds because B_{d} also has full column rank so that the right-hand side is the trivial set while the left-hand side is not trivial because the matrix is fat. In addition, item (b) follows from the facts that the relative degree of $C(zI - A_{\mathsf{d}})^{-1}B_{\mathsf{d}}$ is 1 for almost all sampling periods and that CB_{d} corresponds to the relative degree of the transfer function. It is emphasized again that

$$\text{(c) } \ker \begin{bmatrix} CA_{\mathsf{d}}B_{\mathsf{d}}, & CB_{\mathsf{d}} \end{bmatrix} \text{ is not trivial.}$$

Now, it is ready to design an attack sequence $a_a[i]$. The design proceeds sequentially as a pair, i.e., choosing the sequences $\alpha_m, \beta_m \in \mathbb{R}^2$ such that

$$\begin{bmatrix} a_a[2m] \\ a_a[2m+1] \end{bmatrix} = \beta_m + \alpha_m, \qquad m = 0, 1, \ldots. \tag{2.17}$$

For the case $m = 0$, let $\beta_0 = 0$, and pick α_0 such that

$$\begin{bmatrix} CA_{\mathsf{d}}B_{\mathsf{d}} & CB_{\mathsf{d}} \end{bmatrix}\alpha_0 = 0 \quad \text{and} \quad \begin{bmatrix} B_{\mathsf{d}} & 0 \\ A_{\mathsf{d}}B_{\mathsf{d}} & B_{\mathsf{d}} \end{bmatrix}\alpha_0 \neq 0$$

which is possible by the items (a) and (c). Then, by (2.17) with $m = 0$, it follows from (2.16) that $\tilde{y}[1] = 0$ and at least one of $\tilde{x}(T_a)$ and $\tilde{x}(2T_a)$ is non-zero. Moreover, by choosing α_0 to have a sufficiently large magnitude, the non-zero one can be made arbitrarily large in magnitude. Therefore, both the stealthy property and the disruptive property of the attack are satisfied in the attack period of $2T_a$.

However, it may require too much actuation effort to achieve the level of hazard L_{hazard} in (2.1) within the period of $2T_a$. In order to distribute the effort over time, let us design the attack (2.17) for next steps $m \geq 1$. Since

$$\begin{aligned}
\begin{bmatrix} \tilde{x}((2m+1)T_a) \\ \tilde{x}(2(m+1)T_a) \end{bmatrix} &= \begin{bmatrix} A_{\mathsf{d}}\tilde{x}(2mT_a) \\ A_{\mathsf{d}}^2\tilde{x}(2mT_a) \end{bmatrix} + \begin{bmatrix} B_{\mathsf{d}} & 0 \\ A_{\mathsf{d}}B_{\mathsf{d}} & B_{\mathsf{d}} \end{bmatrix}\begin{bmatrix} a_a[2m] \\ a_a[2m+1] \end{bmatrix} \\
&= \begin{bmatrix} A_{\mathsf{d}}\tilde{x}(2mT_a) \\ A_{\mathsf{d}}^2\tilde{x}(2mT_a) \end{bmatrix} + \begin{bmatrix} B_{\mathsf{d}} & 0 \\ A_{\mathsf{d}}B_{\mathsf{d}} & B_{\mathsf{d}} \end{bmatrix}\beta_m + \begin{bmatrix} B_{\mathsf{d}} & 0 \\ A_{\mathsf{d}}B_{\mathsf{d}} & B_{\mathsf{d}} \end{bmatrix}\alpha_m \qquad (2.18a)\\
\tilde{y}[m+1] = \tilde{y}((m+1)T_s) &= CA_{\mathsf{d}}^2\tilde{x}(2mT_a) + \begin{bmatrix} CA_{\mathsf{d}}B_{\mathsf{d}} & CB_{\mathsf{d}} \end{bmatrix}\begin{bmatrix} a_a[2m] \\ a_a[2m+1] \end{bmatrix} \\
&= CA_{\mathsf{d}}^2\tilde{x}(2mT_a) + \begin{bmatrix} CA_{\mathsf{d}}B_{\mathsf{d}} & CB_{\mathsf{d}} \end{bmatrix}\beta_m + \begin{bmatrix} CA_{\mathsf{d}}B_{\mathsf{d}} & CB_{\mathsf{d}} \end{bmatrix}\alpha_m, \qquad (2.18b)
\end{aligned}$$

the vector β_m is chosen first such that

$$CA_{\mathsf{d}}^2\tilde{x}(2mT_a) + \begin{bmatrix} CA_{\mathsf{d}}B_{\mathsf{d}} & CB_{\mathsf{d}} \end{bmatrix}\beta_m = 0$$

which is possible, thanks to item (b). The role of β_m is to cancel the effect of the previous actuation in (2.18b) in order for the stealthiness of the step m. Then, with β_m at hand, the vector α_m is chosen in $\ker\begin{bmatrix} CA_{\mathsf{d}}B_{\mathsf{d}}, & CB_{\mathsf{d}} \end{bmatrix}$, so that $\tilde{y}[m+1] = 0$ from (2.18b), but not in the kernel of

$$\begin{bmatrix} B_{\mathsf{d}} & 0 \\ A_{\mathsf{d}}B_{\mathsf{d}} & B_{\mathsf{d}} \end{bmatrix}$$

such that the states in (2.18a) can reach the level L_{hazard}. By repeating the procedure for $m = 2, 3, \ldots$, a disruptive stealthy attack can be generated.

It is noted that the attack generation can be performed offline because it does not require any online information such as input and output signals. Therefore, the attack sequence $a_a[i]$ can be computed and stored before the attack initiation.

Example 2.2 Consider a model of X-38 vehicle in [22] which has 11 states, 3 inputs, and 9 outputs. The system admits a multi-rate controller (see [22]), and we suppose that $T_s = 4T_a$ so that the inequality (2.14) holds. Following the procedure of [12], an actuator attack signal is generated offline before the attack initiation. Figure 2.6 illustrates the outcome of the attack. □

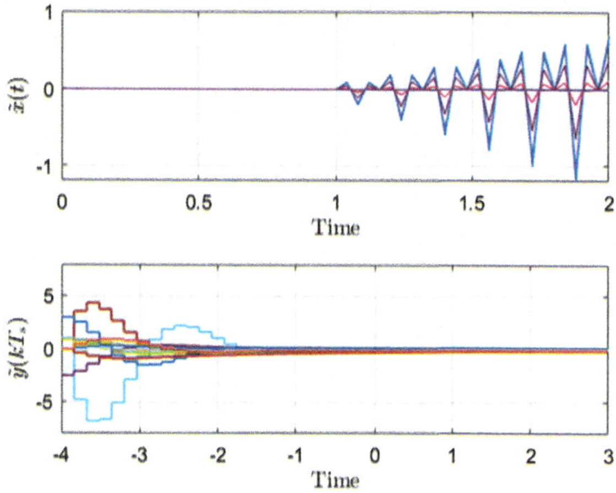

Fig. 2.6 The system reaches its steady state before the time $t_0 = 0$, and enforced-zero-dynamics attack begins at $t_0 = 0$. While the sampled output $\tilde{y}(kT_s)$ remains identically zero, the continuous-time state deviation $\tilde{x}(t)$ is diverging

All of the attacks discussed so far require exact knowledge of the plant. In practice, the exact model of the physical system is hard to obtain not only for attackers but also for control designers. At this moment, a natural question arises: "It seems that these attack policies are too ideal. Is the real control system secure because exact model knowledge is hard to obtain?"

2.2.4 *Robust Zero-Dynamics Attack*

Unfortunately, an exact model of the physical plant is not necessary for generating disruptive stealthy attacks, if the attack policy is designed by employing a robust control methodology. In this subsection, let us investigate this possibility using a particular robust control, called *disturbance observer approach* [23], and generate a disruptive stealthy actuator attack for continuous-time linear systems.

For this, let us write the plant in the normal form again:

$$\begin{aligned} y = x_1, \qquad & \dot{x}_{:} = O^1 x_{:} + 0_1(\phi_{:}^\top x_{:} + \phi_z^\top x_z + b(u + a_a)), \qquad && x_{:} \in \mathbb{R}^r, \\ & \dot{x}_z = S x_z + p x_1, && x_z \in \mathbb{R}^{n-r}. \end{aligned} \tag{2.19}$$

While the proposed attack policy does not need exact values of the system parameters $\phi_{:}$, ϕ_z, b, S, and p, some information regarding the plant is still needed. They are as follows:

- the input and the output of the plant: instead of using less knowledge on the plant model, the attacker relies more on the input and the output information (readers are reminded that the classical zero-dynamics attack does not use the output of the plant);
- the relative degree r, and the sign of the high-frequency gain b, and nominal values $\phi_{:,\mathrm{n}}$, $\phi_{z,\mathrm{n}}$, b_n, S_n, p_n;
- attack-free operating region $\mathscr{X}$ and $\mathscr{Z}$ such that $x_:(t) \in \mathscr{X}$ and $x_z(t) \in \mathscr{Z}$, $\forall t \geq 0$: these are the underlying sets for attack-free operation, and both $\mathscr{X}$ and $\mathscr{Z}$ are assumed to be compact sets; they can be taken conservatively,
- the amount of variation of uncertain parameters: for example, (possibly conservative) interval to which the uncertain parameter b belongs; it is also assumed that the nominal values belong to the intervals.

The above information may be obtained by the attacker from a system identification procedure using the eavesdropped input and output signals, or from leaked information about the design model that was used for constructing the feedback controller. In any case, the amount of information required is not much different from the knowledge to design a robust controller for the uncertain plant. Finally, for technical reasons, we assume the following.

Assumption. The (uncertain) matrix S does not have eigenvalues on the imaginary axis in the complex plane.

To understand the idea of the robust zero-dynamics attack, suppose, for the time being, that the attacker generates the actuator attack signal (instead of (2.5)) by

$$\dot{z}_\mathrm{a} = S_\mathrm{n} z_\mathrm{a} + p_\mathrm{n} y \tag{2.20a}$$

$$a_a^* = \frac{1}{b}\left(-\phi_:^\top x_: - \phi_z^\top x_z - bu + \phi_{:,\mathrm{n}}^\top x_: + \phi_{z,\mathrm{n}}^\top z_\mathrm{a} + b_\mathrm{n} u\right). \tag{2.20b}$$

Of course, the signal a_a^* is not available because many terms on the right-hand side are unknown. This signal a_a^* will be *estimated* by the disturbance observer later. With $a_a = a_a^*$, the plant (2.19) is rewritten (together with (2.20a)) as

$$y = x_1 \qquad \dot{x}_: = O^1 x_: + 0_1(\phi_{:,\mathrm{n}}^\top x_: + \phi_{z,\mathrm{n}}^\top z_\mathrm{a} + b_\mathrm{n} u) \tag{2.21a}$$

$$\dot{z}_\mathrm{a} = S_\mathrm{n} z_\mathrm{a} + p_\mathrm{n} x_1 \tag{2.21b}$$

$$\dot{x}_z = S x_z + p x_1. \tag{2.21c}$$

It is seen from the above equation that all the uncertain parameters are replaced with their nominal values. In particular, the attacker's dynamics (2.21b) replaces the role of the real zero dynamics (2.21c), and disguises as the plant's internal dynamics. The controller interacts with the attacker's dynamics because the state z_a is seen from the output y while the state x_z becomes unobservable, and thus, the real zero dynamics (2.21c) is placed out of the feedback loop.

It is reasonable to assume that the controller robustly stabilizes the uncertain plant whose parameters belong to the uncertainty interval, which also contains the nominal parameters. Then, since the controller stabilizes the plant (2.21) with the replaced nominal parameters, the output y and the state $(x_:, z_a)$ behave as if there is no attack. (This means that the trajectory of $(x_:(t), z_a(t))$ corresponds to $x_{af}(t)$ in (2.2).) In this way, the attack becomes stealthy. On the other hand, if the uncertain plant has an unstable zero so that the matrix S has an unstable eigenvalue (having positive real parts), then the state x_z tends to diverge for *almost all* values of $x_z(t_0)$ where t_0 is the time of attack initiation. In this way, the attack becomes disruptive. It should be noted that the way of obtaining disruptive property is different from that of the zero-dynamics attack in Sect. 2.2.1.

Why does $\mathbf{x_z}$ not diverge for all $\mathbf{x_z(t_0)}$? Even if a system is unstable, there is a measure-zero set of initial conditions from which the solution remains bounded. For example, $x(0) = 0$ for the unstable system $\dot{x} = x$ yields a bounded solution $x(t) = 0$. This is true even when the system has a bounded external input. An example is $\dot{x} = x + \sin t$. With $x(0) = -1/2$, the solution $x(t) = -(1/2)(\sin t + \cos t)$ remains bounded. Now, let us consider (2.21c). Even in the case when all the eigenvalues of S have positive real parts, the initial condition

$$x_z(t_0) = \int_{t_0}^{\infty} e^{-S(t-t_0)} p x_1(t) dt$$

is well-defined because $x_1(t)$ is bounded by the stability of the closed-loop system with the controller. Then, it can be shown that the solution $x_z(t)$ for $t \geq t_0$ is bounded. However, it is not possible to predict such a value $x_z(t_0)$ because the future trajectory of $x_1(t)$, as well as S and p, cannot be known at time t_0. Nevertheless, the set of these particular initial conditions has measure zero, and so, the bounded solution of $x_z(t)$ is not likely to happen in practice.

Now, let us turn to the question of how to estimate a_a^*. In fact, (2.19) can be written as follows with a_a^*:

$$\begin{aligned} \dot{x}_\cdot &= O^1 x_: + 0_1(\phi_:^\top x_: + \phi_z^\top x_z + b(u + a_u)) \\ &= O^1 x_: + 0_1(\phi_{:,n}^\top x_: + \phi_{z,n}^\top z_a + b_n u + b(a_a - a_a^*)). \end{aligned} \quad (2.22)$$

If a_a^* is treated as an unknown disturbance and a_a is the control input, then this is the very situation where the disturbance observer technique can estimate a_a^* with arbitrarily precision. More specifically, if all the initial conditions at $t = 0$ belong to a compact set, and if the disturbance $a_a^*(t)$ is bounded, then, for any $\epsilon > 0$, a disturbance observer and an input a_a can be designed such that

$$\|a_a(t) - a_a^*(t)\| \leq \epsilon, \qquad \forall t \geq \epsilon. \quad (2.23)$$

But, the signal $a_a^*(t)$ is not bounded if $x_z(t)$ is unbounded, which is the goal of the attacker. Therefore, the precise estimation of a_a^* by a_a is not possible for an infinite time horizon. However, this is enough for attackers because, until $x_z(t)$ diverges up to the level of the hazard in (2.1) at time t^* (which takes a finite time duration $t^* - t_0$), the signal $a_a^*(t)$ can be treated as a bounded one and the disturbance observer can be designed to estimate it sufficiently closely.

Finally, the following dynamics generates robust zero-dynamics attack signal a_a, which approximates a_a^* of (2.20), for $t_0 \leq t \leq t^*$:

$$\begin{aligned}
\dot{z}_\mathsf{a} &= S_\mathsf{n} z_\mathsf{a} + p_\mathsf{n} y \\
\dot{\xi}_i &= \xi_{i+1} - \frac{q_{r-i}}{\tau^i}\xi_1 + \frac{q_0}{\tau^r}\frac{1}{b_\mathsf{n}}\left(\phi_{:,\mathsf{n},r-i+1} + \frac{q_{r-i}}{\tau^i}\right) y, \qquad i = 1, \dots, r-1, \\
\dot{\xi}_r &= -\frac{q_0}{\tau^r}\xi_1 + \frac{q_0}{\tau^r}\frac{1}{b_\mathsf{n}}\left(\phi_{:,\mathsf{n},1} + \frac{q_0}{\tau^r}\right) y + \frac{q_0}{\tau^r}\left(\frac{1}{b_\mathsf{n}}\phi_{\mathsf{z},\mathsf{n}}^\top z_\mathsf{a} + u + a_a\right) \qquad (2.24) \\
a_a &= \mathrm{sat}_L\left(\xi_1 - \frac{q_0}{\tau^r}\frac{1}{b_\mathsf{n}} y\right).
\end{aligned}$$

All initial conditions at t_0 are set to zero, assuming that the plant is in the steady state at time t_0.[2] The function sat_L saturates at $\pm L$, and the reason for introducing saturation is that even if the estimation of (2.23) holds for $t \geq t_0 + \epsilon$, nothing can be said for $t_0 \leq t < t_0 + \epsilon$. In order to prevent unnecessary peak of $a_a(t)$ during this period (this peak tends to occur due to the peaking phenomenon [25], and by the peak, the trajectory (2.22) deviates from the attack-free trajectory hindering the stealthiness), we introduce the saturation function. Now, the constants q_i $(i = 0, \dots, r-1)$, the level of saturation L, and the constant τ are design parameters, which are chosen from all the information available to the attacker.

Proposition 2.4 *Let the polynomial $s^{r-1} + q_{r-1}s^{r-2} + \cdots + q_1$ be Hurwitz. If $q_0 > 0$ is chosen sufficiently small, L sufficiently large, and $\tau > 0$ sufficiently small, then the actuator attack generated through (2.24) is stealthy. If, in addition, the matrix S contains at least one eigenvalue having a positive real part, then the attack is disruptive for almost all initial conditions $x_z(t_0)$. Finally, accuracy of estimation of a_a^* by a_a improves as τ gets smaller.*

The proof is omitted due to space limit but can be found in [19], where the details of choosing q_0, L, and τ are presented. For more details on the disturbance observer technique used for (2.24), refer to [2].

[2] As seen in (2.21), the effect of the proposed attack is to replace the real zero dynamics (2.21c) with (2.21b) at time t_0. This is an abrupt change, and so, unless $z_\mathsf{a}(t_0)$ is close to $x_z(t_0)$, some transient response may occur after t_0 and the attack may be detected. To avoid this possibility, the attacker will carefully choose the time t_0 such that the internal state $x_z(t_0)$ is easily guessed, like the steady state at which $z_\mathsf{a}(t_0) = 0$ may suffice.

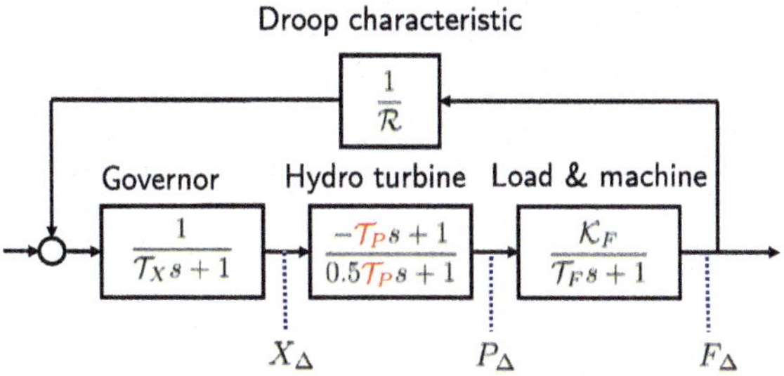

Fig. 2.7 Power-generating plant with hydro turbine. F_Δ: frequency deviation; P_Δ: deviation in generated Power; X_Δ: deviation in governor's valve. The parameter $\mathscr{T}_P$ is uncertain and belongs to the interval [4, 6]. Nominal $\mathscr{T}_{P,\mathrm{n}}$ is taken as 4, and $\mathscr{T}_X = 0.2$, $\mathscr{T}_F = 6$, $\mathscr{K}_F = 1$, and $\mathscr{R} = 0.05$. (©2019 IEEE. Reprinted, with permission, from [19])

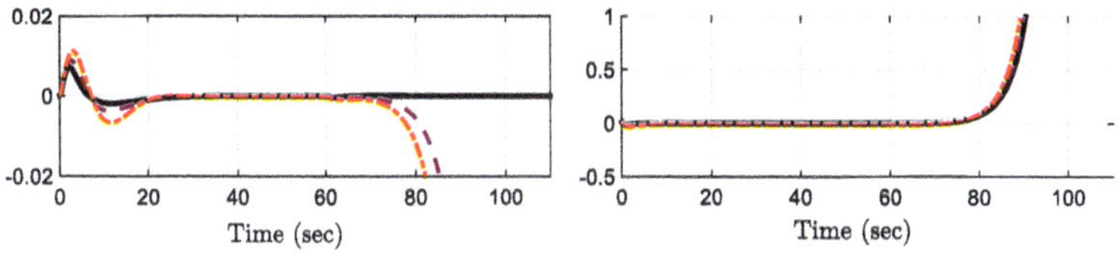

Fig. 2.8 Plots of the output F_Δ (left) and the state X_Δ (right). Attack starts at 60 sec. The black solid is the case when $\mathscr{T}_P = \mathscr{T}_{P,\mathrm{n}} = 4$ while the purple dash means $\mathscr{T}_P = 5$, and the red dash-dot means $\mathscr{T}_P = 6$. With model uncertainty, the attack is easily detectable. (©2016 IEEE. Reprinted, with permission, from [20])

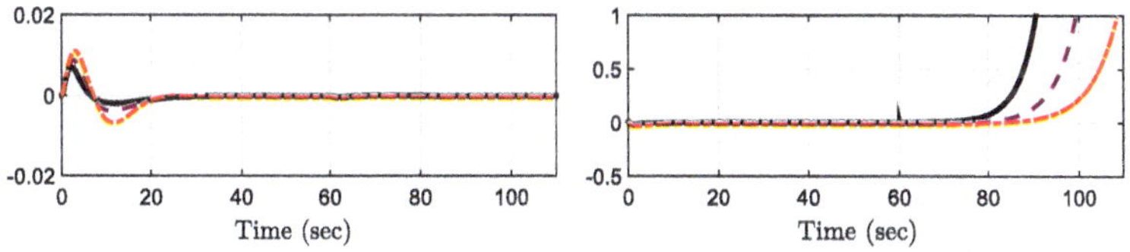

Fig. 2.9 Plots of the output F_Δ (left) and the state X_Δ (right). Attack starts at 60 sec. The black solid is the case when $\mathscr{T}_P = \mathscr{T}_{P,\mathrm{n}} = 4$ while the purple dash means $\mathscr{T}_P = 5$, and the red dash-dot means $\mathscr{T}_P = 6$. In spite of model uncertainty, the attack remains stealthy. (©2016 IEEE. Reprinted, with permission, from [20])

Example 2.3 A non-minimum phase uncertain system depicted in Fig. 2.7 is considered. It is supposed that the plant is controlled by a PID-type controller $C(s) = (1.81s^2 - 18.9s + 0.15)/(0.01s^2 + s)$. The classical zero-dynamics attack in Sect. 2.2.1 with the nominal parameter $\mathscr{T}_{P,\mathrm{n}} = 4$ results in Fig. 2.8. The robust zero-dynamics attack in this section yields Fig. 2.9. For simulation, $q_0 = 1$, $L = 20{,}000$, and $\tau = 0.001$ are used. □

2.2.4.1 Robust Pole-Dynamics Attack

As a dual to robust zero-dynamics attack, robust pole-dynamics attack is also possible. Recall the closed-loop system in (2.7), which is written again as

$$\begin{aligned} u = c_1 \qquad & \dot{c}_{:} = O^1 c_{:} + 0_1(\phi_{:}^{\top} c_{:} + \phi_z^{\top} c_z + b(Cx + a_s)) \\ & \dot{c}_z = S c_z + p c_1 \\ y = Cx \qquad & \dot{x} = Ax + Bc_1. \end{aligned} \tag{2.25}$$

When the plant's parameters A, B, and C have uncertainty and/or the controller's parameters $\phi_{:}$, ϕ_z, b, S, and p are uncertain, the robust pole-dynamics attack can be employed. In particular, the zero-dynamics part in (2.25) and the uncertain parameters can be replaced with their nominal counterparts as

$$\begin{aligned} \dot{c}_{:} &= O^1 c_{:} + 0_1(\phi_{:,\mathrm{n}}^{\top} c_{:} + \phi_{z,\mathrm{n}}^{\top} z_{\mathrm{a},c} + b_{\mathrm{n}} C_{\mathrm{n}} z_{\mathrm{a},x} + b(a_s - a_s^*)) \\ \begin{bmatrix} \dot{z}_{\mathrm{a},c} \\ \dot{z}_{\mathrm{a},x} \end{bmatrix} &= \begin{bmatrix} S_{\mathrm{n}} & 0 \\ 0 & A_{\mathrm{n}} \end{bmatrix} \begin{bmatrix} z_{\mathrm{a},c} \\ z_{\mathrm{a},x} \end{bmatrix} + \begin{bmatrix} p_{\mathrm{n}} \\ B_{\mathrm{n}} \end{bmatrix} c_1, \end{aligned} \tag{2.26}$$

where a_s^* can be estimated by a_s through the disturbance observer as before. For the details, refer to [7].

Up to now, only linear systems are considered as target systems. Is it possible to generate a disruptive stealthy attack for nonlinear systems?

2.2.5 *Zero-Dynamics Attack for Nonlinear Systems*

In fact, nonlinear systems are also vulnerable to disruptive stealthy attacks. When a single-input single-output smooth nonlinear system has a well-defined relative degree r, it can be converted into the nonlinear normal form as

$$\begin{aligned} y = x_1 \qquad & \dot{x}_{:} = O^1 x_{:} + 0_1(f(x_{:}, x_z) + g(x_{:}, x_z)(u + a_a)) \\ & \dot{x}_z = h(x_z, x_{:}) \end{aligned} \tag{2.27}$$

where $x_{:} \in \mathbb{R}^r$ and $x_z \in \mathbb{R}^{n-r}$. For this system, while it is not straightforward to apply the approach of Sect. 2.2.1, the approach of robust zero-dynamics attack can still be applied as long as $g(x_{:}, x_z)$ is a constant function such that $g(x_{:}, x_z) = b$, and $h(x_z, x_{:})$ depends only on (x_z, x_1). With

$$\begin{aligned} \dot{z}_{\mathrm{a}} &= h_{\mathrm{n}}(z_{\mathrm{a}}, y) \\ a_a^* &= \frac{1}{b}(-f(x_{:}, x_z) - bu + f_{\mathrm{n}}(x_{:}, z_{\mathrm{a}}) + b_{\mathrm{n}} u), \end{aligned}$$

the system is rewritten as

$$
\begin{aligned}
y = x_1 \qquad \dot{x}_{:} &= O^1 x_{:} + 0_1(f_{\mathsf{n}}(x_{:}, z_{\mathsf{a}}) + b_{\mathsf{n}} u + b(a_a - a_a^*)) \\
\dot{z}_{\mathsf{a}} &= h_{\mathsf{n}}(z_{\mathsf{a}}, x_1) \\
\dot{x}_z &= h(x_z, x_1).
\end{aligned}
\tag{2.28}
$$

Then, a_a^* can be estimated by a nonlinear disturbance observer studied in [1].

2.3 Disarming Zero-Dynamics Attack

In this section, we discuss how to protect the control system from zero-dynamics attacks when it is implemented in the sampled-data framework. In particular, we consider a SISO LTI continuous-time system

$$
\dot{x}(t) = Ax(t) + Bu(t), \qquad y(t) = Cx(t) \tag{2.29}
$$

that is implemented in the sampled-data framework with

$$
u(t) = u[k] \;\; \forall t \in [kT_s, (k+1)T_s), \qquad y[k] = y(kT_s). \tag{2.30}
$$

Because the attack cannot be detected by monitoring the sampled output $y[k]$, it is generally hard to conceive a countermeasure. Only a few protecting strategies are available in the literature. The idea of changing sensors and actuators is proposed in [26], and the idea of introducing a modulation block before the actuator in the feedback configuration is proposed in [6]. Both methods are intended to modify the system structure under the premise that the changes are kept confidential to the adversaries. Therefore, the security is compromised if the changes are disclosed to adversaries, and even if this is not the case, the sampling-zero-dynamics attack may still be effective when the sampling period T_s is very small. This is because, if T_s is very small, the sampling zeros approach the roots of Euler–Frobenius polynomial, and the Euler–Frobenius polynomial (which is known to the public) does not depend on the particular parameters of the target system (as long as the relative degree of the system is known).

In this section, we present another strategy that can disarm the zero-dynamics attack. The idea is to replace the conventional zero-order hold or the sampler in the plant side with a Generalized Hold (GH) or a Generalized Sampler (GS), respectively.[3] With this change, it is intended that the unstable zeros are relocated to the stable region in the discrete-time domain. Therefore, if the change is not known to the adversaries, then the attack will be detected, and even if the change is known so that the attack is made stealthy, the attack is no more disruptive because all the zeros are stable. Eventually, the adversaries lose the motivation to attack.

---

[3] The generalized hold and sampler have been actively studied in the 90s. For details, refer to [27].

2.3.1 Moving Zeros by Generalized Hold

In this subsection, we introduce the idea of using Generalized Hold (GH), which is based on the work of [11]. Generalized hold [27] is a device that generates the continuous-time input to the plant from a given input sequence $u[k]$ as

$$u_g(t) = h_g(t - kT_s)u[k] \qquad \forall t \in [kT_s, (k+1)T_s) \tag{2.31}$$

where h_g is a function defined on $[0, T_s)$. Recall that if $h_g(t) = 1$ for $[0, T_s)$, then GH recovers zero-order hold. With GH, the sampled-data model of (2.29) becomes

$$x[k+1] = A_\mathrm{d}x[k] + B_g u[k], \quad y[k] = C_\mathrm{d}x[k], \tag{2.32}$$

where $x[k] \in \mathbb{R}^n$ and

$$A_\mathrm{d} = e^{AT_s}, \quad B_g := \int_0^{T_s} e^{A(T_s-\tau)}Bh_g(\tau)d\tau, \quad C_\mathrm{d} = C.$$

The sampled-data transfer function from $u[k]$ to $y[k]$, denoted by $G_\mathrm{d}(z)$, is then given by

$$G_\mathrm{d}(z) = C_\mathrm{d}(zI_n - A_\mathrm{d})^{-1}B_g.$$

Then, in most cases, by a suitable design of h_g, the zeros of $G_\mathrm{d}(z)$ can be assigned to arbitrary locations. To see this, let

$$G_\mathrm{d}^*(z) = k_\mathrm{d}\frac{(z - z_{\mathrm{d},1})\cdots(z - z_{\mathrm{d},n-1})}{\det(zI_n - A_\mathrm{d})} \tag{2.33}$$

be a desired transfer function where $z_{\mathrm{d},1}, \ldots, z_{\mathrm{d},n-1} \in \mathbb{C}$ are the desired zeros (in complex conjugate pairs) and k_d is a desired gain.

Lemma 2.2 *Suppose $(A_\mathrm{d}, C_\mathrm{d})$ is observable.*[4] *Then, there exists a unique $B_g \in \mathbb{R}^n$ such that $G_\mathrm{d}(z) = G_\mathrm{d}^*(z)$.*

Proof Let $C_*(zI_n - A_*)^{-1}B_*$ be a minimal realization of $G_\mathrm{d}^*(z)$. Then, $G_\mathrm{d}(z)$ is identical to $G_\mathrm{d}^*(z)$ if and only if $C_\mathrm{d}A_\mathrm{d}^{k-1}B_g = C_*A_*^{k-1}B_*$ for all $k = 1, 2, \ldots, n$. With $\mathcal{O}_\mathrm{d}$ being the observability matrix of $(A_\mathrm{d}, C_\mathrm{d})$, it is equivalent to

[4] It is well-known that if (A, C) is observable, then $(A_\mathrm{d}, C_\mathrm{d})$ is observable for almost all sampling times T_s.

$$\mathcal{O}_{\mathsf{d}} B_g = \begin{bmatrix} C_{\mathsf{d}} \\ C_{\mathsf{d}} A_{\mathsf{d}} \\ \vdots \\ C_{\mathsf{d}} A_{\mathsf{d}}^{n-1} \end{bmatrix} B_g = \begin{bmatrix} C_* B_* \\ C_* A_* B_* \\ \vdots \\ C_* A_*^{n-1} B_* \end{bmatrix}. \tag{2.34}$$

By the observability, $\mathcal{O}_{\mathsf{d}}$ is invertible. □

With B_g obtained above, it remains to construct $h_g(t)$ such that

$$B_g = \int_0^{T_s} e^{A(T_s-\tau)} B h_g(\tau) d\tau. \tag{2.35}$$

Then, it is immediate to see that such an h_g exists if (A, B) is controllable because the form (2.35) reminds, for $\dot{x} = Ax + Bu$, the problem of driving the state from the origin to the vector B_g at time T_s by the input $u(t) = h_g(t)$. So a solution is given by

$$h_g(t) = B^\top e^{A^\top (T_s - t)} W^{-1}(0, T_s) B_g, \tag{2.36}$$

where $W(0, T_s)$ is the controllability Gramian $W(0, T_s) = \int_0^{T_s} e^{A\tau} B B^\top e^{A^\top \tau} d\tau$. However, this requires a device that can generate continuous-time signals, which may not be very practical.

Alternatively, a piecewise constant function h_g becomes of interest. Let N be the number of subintervals within which $h_g(t)$ is constant, namely

$$h_g(t) = h_i, \quad \frac{(i-1)T_s}{N} \le t < \frac{iT_s}{N}, \quad i = 1, \ldots, N. \tag{2.37}$$

Substituting (2.37) into (2.35), one has

$$B_g = \sum_{i=1}^{N} \int_{\frac{(i-1)T_s}{N}}^{\frac{iT_s}{N}} e^{A(T_s-\tau)} B d\tau \, h_i,$$

which can be simply written, with $\bar{h} := [h_1, \ldots, h_N]^\top$, as

$$B_g = \left[A_{\mathsf{d},N}^{N-1} B_{\mathsf{d},N}, \cdots, A_{\mathsf{d},N} B_{\mathsf{d},N}, B_{\mathsf{d},N} \right] \bar{h} =: M_B \bar{h}, \tag{2.38}$$

where $A_{\mathsf{d},N} = e^{AT_s/N}$, $B_{\mathsf{d},N} = \int_0^{T_s/N} e^{A(T_s/N-\tau)} B d\tau$, and $M_B \in \mathbb{R}^{n \times N}$.

Therefore, we obtain the following result.

Lemma 2.3 *If B_g belongs to the range space of M_B, then the vector $\bar{h}$ exists. If $(A_{\mathsf{d},N}, B_{\mathsf{d},N})$ is controllable (which is the case for almost all T_s when (A, B) is controllable), then the vector $\bar{h}$ exists for any B_g with $N = n$, and $\bar{h} = M_B^{-1} B_g$.*

2.3.2 Optimal Design of Generalized Hold

The approach of moving the system zeros to the stable region by GH is effective in disarming the zero-dynamics attack. However, the response of the resulting system with GH is different from that with ZOH. In other words, the closed-loop system performance is affected by designing the GH and it may degrade from the original design. The question "how can we design GH so that the performance degradation of the closed-loop system (from that with ZOH) is minimized?" naturally follows, which is the topic of this subsection.

In order to design GH following the procedure developed in the previous subsection, one needs to choose the desired zeros first, and compute B_g and then h_g in order. But, then, the values of $h_g(t)$ may happen to be very different from those of the ZOH (which are 1). To overcome this drawback, by recalling that our interest is not to assign the zeros at specific locations but to make the system have stable discrete-time zeros, we formulate an optimization problem to minimize the difference between $h_g(t)$ and 1, under the constraints that the zeros are located inside the unit circle and that the integrations of $h_g(t)$ and 1 during one sampling period are equal.

For building the stability constraint, we note that the realization of $G_{\mathsf{d}}^*(z)$ of (2.33) in the controllable canonical form is given by

$$\bar{x}[k+1] = \left[\begin{array}{c|c} 0_{n-1} & I_{n-1} \\ \hline -d_0 & \cdots \; -d_{n-1} \end{array}\right] \bar{x}[k] + \begin{bmatrix} 0_{n-1} \\ 1 \end{bmatrix} u[k] =: A_* \bar{x}[k] + B_* u[k]$$
$$y[k] = \begin{bmatrix} c_0 & \cdots & c_{n-2} & c_{n-1} \end{bmatrix} \bar{x}[k] =: C_* \bar{x}[k],$$

where the constants $c_0, \ldots, c_{n-1}, d_0, \ldots, d_{n-1}$ are determined from the relations

$$\det(zI_n - A_{\mathsf{d}}) = z^n + d_{n-1}z^{n-1} + \cdots + d_0$$
$$k_{\mathsf{d}} \prod_{i=1}^{n-1} (z - z_{\mathsf{d},i}) = c_{n-1}z^{n-1} + c_{n-2}z^{n-2} + \cdots + c_0. \tag{2.39}$$

Then, it follows from (2.34) and (2.38) that

$$\mathscr{O}_{\mathsf{d}} M_B \bar{h} = \mathscr{O}_{\mathsf{d}} B_g = \begin{bmatrix} C_* B_* \\ \vdots \\ C_* A_*^{n-1} B_* \end{bmatrix} = \begin{bmatrix} B_*^\top \\ \vdots \\ B_*^\top (A_*^\top)^{n-1} \end{bmatrix} C_*^\top =: \mathscr{C}_*^\top C_*^\top, \tag{2.40}$$

where $\mathscr{C}_*$ is the controllability matrix of (A_*, B_*), and we used the fact that $C_* A_*^{i-1} B_*$ is scalar in the derivation.

Remark 2.2 Suppose that we seek for a Schur stable polynomial of order q; i.e., $\alpha_q z^q + \alpha_{q-1} z^{q-1} + \cdots + \alpha_0$. By utilizing the LMI condition for a strict positive real system, an LMI specification for the search can be formulated by

$$
\begin{aligned}
&\text{find } \tilde{C} = [\alpha_0, \alpha_1, \dots, \alpha_{q-1}],\ \tilde{D} = \alpha_q > 0 \text{ and } \tilde{P} > 0 \\
&\text{such that} \quad \begin{bmatrix} \tilde{A}^\top \tilde{P} \tilde{A} - \tilde{P}, & \tilde{A}^\top \tilde{P} \tilde{B} - \tilde{C}^\top \\ \tilde{B}^\top \tilde{P} \tilde{A} - \tilde{C}, & \tilde{B}^\top \tilde{P} \tilde{B} - 2\tilde{D} \end{bmatrix} \le 0,
\end{aligned}
$$

where

$$
\tilde{A} = \begin{bmatrix} 0 & 1 & 0 & \cdots & 0 \\ 0 & 0 & 1 & \cdots & 0 \\ \vdots & \vdots & \vdots & \ddots & \vdots \\ 0 & 0 & 0 & \cdots & 1 \\ 0 & 0 & 0 & \cdots & 0 \end{bmatrix} \in \mathbb{R}^{q \times q}, \quad \tilde{B} = \begin{bmatrix} 0 \\ 0 \\ \vdots \\ 0 \\ 1 \end{bmatrix} \in \mathbb{R}^q.
$$

The underlying idea is to ask the solver to search for $\tilde{C}$ and $\tilde{D}$ such that

$$
\tilde{C}(zI - \tilde{A})^{-1}\tilde{B} + \tilde{D} = \frac{\alpha_q z^q + \cdots + \alpha_0}{z^q}
$$

becomes strictly positive real (which is already stable since $\tilde{A}$ is Schur) [5]. □

Based on the idea of Remark 2, an optimization problem is formulated as

$$
\underset{\bar{h} \in \mathbb{R}^N,\ \tilde{P} \in \mathbb{R}^{(n-1)\times(n-1)}}{\text{minimize}} \quad \|\bar{h} - 1_N\|^2 \tag{2.41a}
$$

$$
\text{subject to} \quad \begin{bmatrix} \tilde{A}^\top \tilde{P} \tilde{A} - \tilde{P}, & \tilde{A}^\top \tilde{P} \tilde{B} \\ \tilde{B}^\top \tilde{P} \tilde{A}, & \tilde{B}^\top \tilde{P} \tilde{B} \end{bmatrix} - (\mathscr{C}_*^{-\top} \mathscr{O}_{\mathrm{d}} M_B \bar{h}) e_n^\top - e_n (\mathscr{C}_*^{-\top} \mathscr{O}_{\mathrm{d}} M_B \bar{h})^\top \le 0 \tag{2.41b}
$$

$$
\text{and} \quad \tilde{P} > 0, \tag{2.41c}
$$

$$
\text{and} \quad 1_N^\top \bar{h} = N, \tag{2.41d}
$$

where $1_N \in \mathbb{R}^N$ is the vector with all elements being 1, and $e_n \in \mathbb{R}^n$ is the elementary vector whose nth element is 1 while all others are zero. Indeed, the column vector $(\mathscr{C}_*^{-\top} \mathscr{O}_{\mathrm{d}} M_B \bar{h})$ is nothing but $[c_0, c_1, \dots, c_{n-1}]^\top$ due to (2.39) and (2.40). The task of disarming the attack is achieved mainly by (2.41b) and (2.41c), and thus, the cost function and the constraint (2.41d) can be modified to obtain different $\bar{h}$.

2.3.3 *Moving Zeros by Generalized Sampler*

In this subsection, it is shown that by using a Generalized Sampler (GS), the zeros of the discrete-time representation of the plant can also be placed inside the unit circle. Zero-dynamics attack may remain stealthy, but since all attack signals diminish, it becomes not disruptive. Along with the approach of using GH, the use of GS disarms the zero-dynamics attack. This subsection is based on the work of [10].

Generalized sampler [27] is a device that generates the output sample from the continuous-time output of the plant as

$$y_g[k] = \int_0^{T_s} h_g(t) y(t + (k-1)T_s) dt \tag{2.42}$$

where h_g is a (generalized) function defined on $(0, T_s]$. The conventional sampler $y[k] = y(kT_s)$ can be considered as a GS with $h_g(t) = \delta(t - T_s)$. Recalling that[5] $y(t + (k-1)T_s) = Ce^{At}x[k-1] + C\int_0^t e^{A\tau}Bd\tau\, u[k-1]$ for $0 < t \le T_s$, it is seen that

$$\begin{aligned} y_g[k] &= \left(\int_0^{T_s} h_g(t) C e^{At} dt\right) x[k-1] + \left(\int_0^{T_s} h_g(t) C \int_0^t e^{A\tau} B d\tau dt\right) u[k-1] \\ &=: C_g\, x[k-1] + D_g\, u[k-1]. \end{aligned} \tag{2.43}$$

With $x[k+1] = e^{AT_s}x[k] + \int_0^{T_s} e^{A\tau}Bd\tau u[k] =: A_\mathsf{d}x[k] + B_\mathsf{d}u[k]$, the transfer function from $u[k]$ to $y_g[k+1]$ is given by $C_g(zI_n - A_\mathsf{d})^{-1}B_\mathsf{d} + D_g$, and therefore, the transfer function from $u[k]$ to $y_g[k]$ becomes

$$G_\mathsf{d}(z) = z^{-1}(C_g(zI_n - A_\mathsf{d})^{-1}B_\mathsf{d} + D_g) \tag{2.44}$$

which has dimension $n + 1$ (rather than n) due to the one-step delay. Then, in most cases, by a suitable design of h_g, the zeros of $G_\mathsf{d}(z)$ can be assigned to arbitrary locations. To see this, let

$$G_\mathsf{d}^*(z) = k_\mathsf{d} z^{-1} \frac{(z - z_{\mathsf{d},1}) \cdots (z - z_{\mathsf{d},n})}{\det(zI_n - A_\mathsf{d})} \tag{2.45}$$

be a desired transfer function where $z_{\mathsf{d},1}, \ldots, z_{\mathsf{d},n} \in \mathbb{C}$ are the desired zeros (in complex conjugate pairs) and k_d is a desired gain.

Lemma 2.4 *Suppose $(A_\mathsf{d}, B_\mathsf{d})$ is controllable.*[6] *Then, there exists a unique pair $C_g \in \mathbb{R}^{1\times n}$ and $D_g \in \mathbb{R}$ such that $G_\mathsf{d}(z) = G_\mathsf{d}^*(z)$.*

Proof Let $C_*(zI_n - A_*)^{-1}B_* + D_*$ is a minimal realization of $zG_\mathsf{d}^*(z)$. Then, $zG_\mathsf{d}(z)$ is identical to $zG_\mathsf{d}^*(z)$ if and only if $D_g = D_*$ and $C_g A_\mathsf{d}^{k-1} B_\mathsf{d} = C_* A_*^{k-1} B_*$ for all $k = 1, 2, \ldots, n$. With $\mathscr{C}_\mathsf{d}$ being the controllability matrix of $(A_\mathsf{d}, B_\mathsf{d})$, it is equivalent to

[5] Note that $\int_0^t e^{A(t-\tau)}Bd\tau = \int_0^t e^{A\tau}Bd\tau$.

[6] It is well-known that if (A, B) is controllable, then $(A_\mathsf{d}, B_\mathsf{d})$ is controllable for almost all sampling times T_s.

$$\begin{aligned} C_g \mathscr{C}_{\mathsf{d}} &= C_g \left[B_{\mathsf{d}}, A_{\mathsf{d}} B_{\mathsf{d}}, \ldots, A_{\mathsf{d}}^{n-1} B_{\mathsf{d}}\right] \\ &= [C_* B_*, \ C_* A_* B_*, \ \ldots, \ C_* A_*^{n-1} B_*] \ \text{ and } \ D_g = D_* = k_{\mathsf{d}}. \end{aligned} \tag{2.46}$$

By the controllability, $\mathscr{C}_{\mathsf{d}}$ is invertible. □

From now on, let us consider the case when $h_g(t) = \sum_{i=1}^{N} w_i \delta(t - iT_s/N)$, where N is a positive integer and w_i's are constant weights, so that

$$y_g[k] = \sum_{i=1}^{N} w_i y\left(\frac{i}{N} T_s + (k-1)T_s\right). \tag{2.47}$$

While (2.42) requires continuous measurements over the interval, only finite number of measurements are asked in (2.47), which is practical.[7]

Now let us look at the relation between the pair (C_g, D_g) and $\bar{w} := [w_1, \ldots, w_N]^\top$. For this, note that, for $i = 1, \ldots, N$,

$$\begin{aligned} x\left(\frac{i}{N} T_s + (k-1)T_s\right) &= e^{A\frac{iT_s}{N}} x[k-1] + \int_0^{\frac{iT_s}{N}} e^{A(\frac{iT_s}{N} - \tau)} B d\tau \; u[k-1] \\ &= A_{\mathsf{d},N}^{i} \, x[k-1] + \sum_{j=1}^{i} A_{\mathsf{d},N}^{j-1} B_{\mathsf{d},N} \, u[k-1], \end{aligned} \tag{2.48}$$

where $A_{\mathsf{d},N} = e^{AT_s/N}$ and $B_{\mathsf{d},N} = \int_0^{T_s/N} e^{A(T_s/N - \tau)} B d\tau = \int_0^{T_s/N} e^{A\tau} B d\tau$. Then, by comparing (2.47) and (2.48) with (2.43), it follows that

$$C_g = \bar{w}^\top \begin{bmatrix} C A_{\mathsf{d},N} \\ C A_{\mathsf{d},N}^2 \\ \vdots \\ C A_{\mathsf{d},N}^N \end{bmatrix} =: \bar{w}^\top M_C, \quad D_g = \bar{w}^\top \begin{bmatrix} C B_{\mathsf{d},N} \\ C A_{\mathsf{d},N} B_{\mathsf{d},N} + C B_{\mathsf{d},N} \\ \vdots \\ C \sum_{j=1}^{N} A_{\mathsf{d},N}^{j-1} B_{\mathsf{d},N} \end{bmatrix} =: \bar{w}^\top M_D, \tag{2.49}$$

where $M_C \in \mathbb{R}^{N \times n}$ and $M_D \in \mathbb{R}^{N \times 1}$. Therefore, we obtain the following result.

Lemma 2.5 *If $[C_g, D_g] \in \mathbb{R}^{1 \times (n+1)}$ belongs to the row space of $[M_C, M_D] \in \mathbb{R}^{N \times (n+1)}$, then the vector $\bar{w}$ exists as a solution to*

$$\bar{w}^\top [M_C, M_D] = [C_g, D_g]. \tag{2.50}$$

If the extended system

$$\tilde{x}[k+1] = \begin{bmatrix} A_{\mathsf{d},N} & B_{\mathsf{d},N} \\ 0 & 1 \end{bmatrix} \tilde{x}[k], \quad \tilde{y}[k] = \left[C A_{\mathsf{d},N}, \; C B_{\mathsf{d},N}\right] \tilde{x}[k],$$

[7] A similar strategy was presented in [17] where a multi-rate sampler is employed for attack detection. However, all samples $y(iT_s/N + (k-1)T_s)$, $i = 1, \ldots, N$, are transmitted to the controller for attack detection.

where $\tilde{x} \in \mathbb{R}^{n+1}$ is observable, then the vector $\bar{w}$ exists for any $[C_g, D_g]$ with $N = n+1$, and is obtained by $\bar{w}^\top = [C_g, D_g][M_C, M_D]^{-1}$.

Proof The first claim is straightforward, and the second claim follows from the fact that the observability matrix of the extended system is $[M_C, M_D]$. □

2.3.4 Optimal Design of Generalized Sampler

Unlike the undesirable inter-sample behavior caused by the generalized hold, the generalized sampler does not affect the inter-sample behavior of the continuous-time system. Instead, the output $y_g[k]$ is different from the conventional output sample $y[k]$, which may not be desirable for monitoring or for other purposes. To mitigate this difference, an optimization problem can be formulated.[8] In particular, we want to reduce $\|y_g[k] - y[k]\|$ under the constraints that the zeros are located inside the unit circle and that the sum of the weights is unity, i.e., $1_N^\top \bar{w} = 1$. Reducing the difference $\|y_g[k] - y[k]\|$ is related to the selection of $\bar{w}$ because

$$y_g[k] = \bar{w}^\top [M_C, M_D] \begin{bmatrix} x[k-1] \\ u[k-1] \end{bmatrix} \quad \text{and} \quad y[k] = e_N^\top [M_C, M_D] \begin{bmatrix} x[k-1] \\ u[k-1] \end{bmatrix}$$

where $e_N = [0, \ldots, 0, 1]^\top \in \mathbb{R}^N$, in which the left equation follows from (2.43) and (2.49), and the right equation is from (2.47) by observing that (2.47) becomes $y[k] = y(kT_s)$ when $w_N = 1$ and $w_i = 0$ for all $i \neq N$.

Now, we need to relate $\bar{w}$ with the stability of the zero polynomial in (2.45). This is achieved by realizing $zG_{\mathsf{d}}^*(z)$ in the controllable canonical form as

$$\begin{aligned} \bar{x}[k+1] &= \left[\begin{array}{c|c} 0_{n-1} & I_{n-1} \\ \hline -d_0 \ \cdots & -d_{n-1} \end{array} \right] \bar{x}[k] + \begin{bmatrix} 0_{n-1} \\ 1 \end{bmatrix} u[k] =: A_* \bar{x}[k] + B_* u[k] \\ y_g[k+1] &= \begin{bmatrix} c_0 \cdots c_{n-2} \ c_{n-1} \end{bmatrix} \bar{x}[k] + k_{\mathsf{d}} u[k] =: C_* \bar{x}[k] + D_* u[k] \end{aligned}$$

where $D_* = k_{\mathsf{d}}$ and the constants $c_0, \ldots, c_{n-1}, d_0, \ldots, d_{n-1}$ are determined from the relations

$$\begin{aligned} \det(zI_n - A_{\mathsf{d}}) &= z^n + d_{n-1}z^{n-1} + \cdots + d_0 \\ k_{\mathsf{d}} \left(\prod_{i=1}^{n} (z - z_{\mathsf{d},i}) - \det(zI_n - A_{\mathsf{d}}) \right) &= c_{n-1}z^{n-1} + c_{n-2}z^{n-2} + \cdots + c_0. \end{aligned} \tag{2.51}$$

With $C_* = [c_0, \ldots, c_{n-1}]$ and $D_* = k_{\mathsf{d}}$, it follows from (2.50) and (2.46) that

[8] This subsection is a brief summary of the contribution of [9].

$$\begin{aligned}\bar{w}^\top M_C &= C_g = \left[C_* B_*, \ \ldots, \ C_* A_*^{n-1} B_*\right] \mathscr{C}_{\mathsf{d}}^{-1} = C_* \left[B_*, \ \ldots, \ A_*^{n-1} B_*\right] \mathscr{C}_{\mathsf{d}}^{-1} \\ &=: C_* \mathscr{C}_* \mathscr{C}_{\mathsf{d}}^{-1}, \\ \bar{w}^\top M_D &= D_g = D_* = k_{\mathsf{d}}.\end{aligned}$$

Finally, the convex optimization problem is given by

$$\underset{\bar{w}\in\mathbb{R}^N,\ \tilde{P}\in\mathbb{R}^{n\times n}}{\text{minimize}} \quad \|[M_C, M_D]^\top (\bar{w} - e_N)\|^2 \tag{2.52a}$$

$$\text{subject to} \quad \begin{bmatrix} \tilde{A}^\top \tilde{P} \tilde{A} - \tilde{P}, & \tilde{A}^\top \tilde{P} \tilde{B} - \tilde{C}^\top \\ \tilde{B}^\top \tilde{P} \tilde{A} - \tilde{C}, & \tilde{B}^\top \tilde{P} \tilde{B} - 2\tilde{D} \end{bmatrix} \le 0, \quad \tilde{P} > 0, \tag{2.52b}$$

$$\text{and} \quad \tilde{C} = \mathscr{C}_*^{-\top} \mathscr{C}_{\mathsf{d}}^\top M_C^\top \bar{w} + (M_D^\top \bar{w}) \bar{d}, \quad \tilde{D} = M_D^\top \bar{w} \tag{2.52c}$$

$$\text{and} \quad 1_N^\top \bar{w} = 1 \tag{2.52d}$$

where $\bar{d} := [d_0, d_1, \ldots, d_{n-1}]^\top$. The constraints (2.52b) and (2.52c) are for stability of the zero polynomial $k_{\mathsf{d}} \prod_{i=1}^n (z - z_{\mathsf{d},i})$, derived from (2.51) and Remark 2. The task of disarming the attack is achieved mainly by (2.52b) and (2.52c), and thus, the cost function and the constraint (2.52d) can be modified to obtain different $\bar{w}$.

2.4 Conclusion

This chapter introduces an actuator attack that is both stealthy and disruptive. In particular, the attack exploits the zero dynamics of the system, which is either from intrinsic zeros or from sampling zeros. Also presented is the robust zero-dynamics attack that is both disruptive and stealthy but the design of which does not require exact knowledge of the target system. This implies that the target control system may be in greater danger than one would imagine. The attack extends to the case of nonlinear systems. As a dual problem of the zero-dynamics attack, two sensor attacks are briefly introduced as well, which are referred to as pole-dynamics attack and robust pole-dynamics attack, respectively. These are attacks on sensors that exploit the plant pole dynamics. Finally, two methods are introduced that render the zero-dynamics attack to be little of a threat. It is shown that by designing a generalized hold or sample, unstable zeros move to a stable region in the complex plane. Thus, the attacks designed for the zeros are no longer disruptive. These kinds of solutions are effective compared to other countermeasures against zero-dynamics attack, but require a redesign of feedback controllers because the discrete-time transfer function is changed by the generalized hold or sampler. In addition, the robustness of the designed hold or sampler against the uncertainty of the plant needs to be investigated as the issue has been raised in, e.g., [3].

Acknowledgements The authors are grateful to Hyuntae Kim at Seoul National University for his idea of Remark 2. This work was supported by Institute for Information & communications Technology Promotion grant funded by MSIT, the Korean government (2014-0-00065, Resilient Cyber-Physical Systems Research).

References

1. Back, J., Shim, H.: Adding robustness to nominal output-feedback controllers for uncertain nonlinear systems: a nonlinear version of disturbance observer. Automatica **44**(10), 2528–2537 (2008)
2. Back, J., Shim, H.: Reduced-order implementation of disturbance observers for robust tracking of non-linear systems. IET Control Theory Appl. **8**(17), 1940–1948 (2014)
3. Freudenberg, J.S., Middleton, R.H., Braslavsky, J.H.: Robustness of zero shifting via generalized sampled-data hold functions. IEEE Trans. Autom. Control **42**(12), 1681–1692 (1997)
4. Giles, M.: Triton is the world's most murderous malware, and it's spreading. MIT Technology Review, March 5 (2019)
5. Hitz, B.E., Anderson, B.D.O.: Discrete positive-real functions and their application to system stability. Proc. IEE **116**(1), 153–155 (1969)
6. Hoehn, A., Zhang, P.: Detection of covert attacks and zero dynamics attacks in cyber-physical systems. In: Proceedings of American Control Conference, pp. 302–307 (2016)
7. Jeon, H., Aum, S., Shim, H., Eun, Y.: Resilient state estimation for control systems using multiple observers and median operation. Math. Probl. Eng. **2016**, 3750264 (2016)
8. Khalil, H.K.: Nonlinear Control. Pearson Higher Ed (2014)
9. Kim, D., Ryu, K., Back, J.: Zero assignment via generalized sampler: a countermeasure against zero-dynamics attack. submitted
10. Kim, D., Ryu, K., Back, J.: Security enhancement of sampled-data systems: zero assignment via generalized sampler. In: Proceedings of IFAC World Congress (2020)
11. Kim, J., Back, J., Park, G., Lee, C., Shim, H., Voulgaris, P.G.: Neutralizing zero dynamics attack on sampled-data systems via generalized holds. Automatica **113** (2020)
12. Kim, J., Park, G., Shim, H., Eun, Y.: Masking attack for sampled-data systems via input redundancy. IET Control Theory Appl. **13**(14), 2300–2308 (2019)
13. Langner, R.: Stuxnet: dissecting a cyberwarfare weapon. IEEE Secur. Priv. **9**(3), 49–51 (2011)
14. Lee, R.M., Assante, M.J., Conway, T.: Analysis of the cyber attack on the Ukrainian power grid. SANS Industrial Control Systems, Washington, DC. USA, Technical Report (2016)
15. Lee, C., Shim, H., Eun, Y.: On redundant observability: from security index to attack detection and resilient state estimation. IEEE Trans. Autom. Control **64**(2), 775–782 (2019)
16. Lee, J.G., Kim, J., Shim, H.: Fully distributed resilient state estimation based on distributed median solver. IEEE Trans. Autom. Control **65**(9), 3935–3942 (2020)
17. Naghnaeian, M., Hirzallah, N.H., Voulgaris, P.G.: Security via multirate control in cyber-physical systems. Syst. Control Lett. **124**, 12–18 (2019)
18. Park, G., Lee, C., Shim, H.: On stealthiness of zero-dynamics attacks against uncertain nonlinear systems: a case study with quadruple-tank process. In: Proceedings of International Symposium on Mathematical Theory of Networks and Systems (MTNS), pp. 10–17 (2018)
19. Park, G., Lee, C., Shim, H., Eun, Y., Johansson, K.H.: Stealthy adversaries against uncertain cyber-physical systems: threat of robust zero-dynamics attack. IEEE Trans. Autom. Control **64**(12), 4907–4919 (2019)
20. Park, G., Shim, H., Lee, C., Eun, Y., Johansson, K.H.: When adversary encounters uncertain cyber-physical systems: Robust zero-dynamics attack with disclosure resources. In: Proceedings of IEEE 55th Conference on Decision and Control, pp. 5085–5090 (2016)
21. Shane, S., Sanger, D.E.: Drone crash in Iran reveals secret US surveillance effort. The New York Times, December 8 (2011)

22. Shieh, L.S., Wang, W.M., Bain, J., Sunkel, J.W.: Design of lifted dual-rate digital controllers for X-38 vehicle. J. Guid. Control Dyn. **23**(4), 629–639 (2000)
23. Shim, H., Park, G., Joo, Y., Back, J., Jo, N.H.: Yet another tutorial of disturbance observer: robust stabilization and recovery of nominal performance. Control Theory Technol. **14**(3), 237–249 (2016)
24. Slay, J., Miller, M.: Lessons learned from the Maroochy water breach. Crit. Infrastruct. Prot. **253**, 73–82 (2007)
25. Sussmann, H., Kokotovic, P.: The peaking phenomenon and the global stabilization of nonlinear systems. IEEE Trans. Autom. Control **36**(4), 424–440 (1991)
26. Teixeira, A., Shames, I., Sandberg, H., Johansson, K.H.: Revealing stealthy attacks in control systems. In: Proceedings of 50th Annual Allerton Conference on Communication, Control, and Computing, pp. 1806–1813 (2012)
27. Yuz, J.I., Goodwin, G.C.: Sampled-Data Models for Linear and Nonlinear Systems. Springer (2014)

Chapter 3
Secure Networked Control Under Jamming Attacks: An SINR-Based Approach

Ahmet Cetinkaya, Hideaki Ishii, and Tomohisa Hayakawa

Abstract Ensuring cybersecurity in networked control systems is a big challenge, especially when wireless channels are used for transmissions of measurement and control data. These channels are subject to a variety of critical threats, among which jamming attacks appear to be one of the easiest to achieve from the perspective of an attacker. Here we look at the problem of controlling a discrete-time linear system over an insecure wireless channel that faces jamming attacks. We consider a channel model based on Signal-to-Interference-plus-Noise Ratio (SINR), where the likelihood of transmission failures at each time can differ depending on the power of the interference signal emitted by a jamming attacker at that time. We investigate state-dependent jamming attacks, where the attacker uses the information of the system state and the dynamics to efficiently change the interference power to cause instability. We analyze the stability of the networked control system by investigating the properties of a martingale that depends on the transmission failure indicator and the interference power process. We then obtain sufficient stability conditions indicating that closed-loop stability can be guaranteed if the attacker has energy constraints and the average jamming interference power has a sufficiently small upper bound. The effect of state-dependent jamming and the utility of our analysis approach is illustrated through an attack strategy with rolling-horizon optimization, where the attacker decides the interference power based on maximizing a utility function that involves the predicted future states given the present state information.

A. Cetinkaya
National Institute of Informatics, Tokyo 101-8430, Japan
e-mail: cetinkaya@nii.ac.jp

H. Ishii (✉) · T. Hayakawa
Tokyo Institute of Technology, Yokohama 226-8502, Japan
e-mail: ishii@c.titech.ac.jp

T. Hayakawa
e-mail: hayakawa@sc.e.titech.ac.jp

H. Ishii and Q. Zhu (eds.), *Security and Resilience of Control Systems*,
Lecture Notes in Control and Information Sciences 489,
https://doi.org/10.1007/978-3-030-83236-0_3

3.1 Introduction

Wireless communication technologies are essential building blocks of the Internet of Things (IoT) and they are becoming more widespread in cyber-physical system applications. In particular, vehicle platoons, smart factories, and smart cities all utilize wireless channels. Any security threat against the communication infrastructure in these applications risks the delivery of services that are essential to the society.

Wireless channels can also be a major source of vulnerability in networked control operation, as discussed thoroughly in the recent literature on control system security [12, 42]. These channels carry important data concerning system measurements and control inputs, and yet they can be easily targeted.

As categorized in recent survey articles [13, 17, 33], there are several ways an attacker can exploit communication channels to target a networked control system. First of all, attackers who gain control of a channel can change the content of data packets. Moreover, they can inject false measurement and control data in the system [19, 36]. In certain cases, where the attacker has detailed information of the system dynamics, attacker can conduct data manipulations in a way that is impossible to detect.

In systems that implement cryptographic security measures, attackers may fail to change packet contents without being detected. However, such systems may still be vulnerable to Denial-of-Service (DoS) attacks, which cause failures in the delivery of packets. There are several ways a DoS attack can take place. For example, on multi-hop ad hoc networks, DoS attacks can be initiated by malicious packet-dropping nodes in the network. The works [27, 28, 34] provide a discussion on the security of multi-hop networks and methods to mitigate DoS attacks.

On wireless networks, DoS attacks can also be performed through a jamming device. An attacker who possesses a jamming device can easily target mobile communications and WiFi [39]. For example, [39] mentions that the IEEE 802.11 wireless communication protocols can be targeted by off-the-shelf hardware with wireless transmitters. An adversary following the protocol can attack the medium access control (MAC) layer and cause packet collisions [21]. On the other hand, a jamming attacker may also choose to perform *physical layer attacks*. In that case, the attacker doesn't even need to follow the protocol. By releasing interference signals, the attacker can cause a reduction in the Signal-to-Interference-plus-Noise Ratio (SINR), which is the ratio of the power of the original data transmission signal to the attacker's interference power summed with the power of the channel noise. When the SINR is too small, the receiver may fail to detect packets transmitted from a sender [21]. Packets that face jamming attacks may get corrupted and cyclic redundancy checks (CRC) performed by the receiver are unsuccessful, resulting in transmission failures.

Recently, networked control systems under jamming and other types of DoS attacks have been investigated in quite a few works including [6, 15, 16, 20, 30, 31, 45, 48]. A detailed overview of these works is provided in [10]. These works can be considered to be complementary to the more classical networked control literature [3,

24–26, 29] that deal with packet transmission failures due to *nonmalicious* reasons (e.g., channel unreliability and congestion). The difference in these two lines of works is the way that the occurrences of the failures are modeled. Specifically, nonmalicious transmission failures are typically modeled in the literature by Bernoulli processes or Markov processes. Such models may not be effective for describing transmission failures due to jamming and other DoS attacks. When characterizing DoS-related failures, one needs to account for the uncertainty in the attacker's strategy. The strategy of the attacker and, in particular, the timing of attacks cannot be exactly known a priori. It is possible to consider the worst-case attacks that maximize the damage on the system, and in that case, a process that describes the occurrences of transmission failures may not even possess the Markov property. There are a few different approaches for modeling these types of attacks in networked control systems.

The work [45] proposed models where the attack strategies are composed of jamming and sleeping cycles with unknown periods. Moreover, [15, 16, 20] considered a more general setting that allows the timing of the attacks to be arbitrary as long as the frequency and the duration of attacks satisfy certain bounds. This general modeling approach was also used in analyzing DoS attacks in multi-agent systems (see, e.g., [11, 43]). A discrete-time version of the model in [15] was used in [6] to characterize the timing of jamming attacks for the scenarios where the failures can be caused either by nonmalicious or malicious reasons. It is shown in [6] that a probabilistic model can generalize the models of both random packet losses described with general Markov processes as well as attacks with unknown timings and bounded average durations. This probabilistic model allows uncertainty in the occurrences of failures, and furthermore, the failure probabilities at each transmission attempt need not be specified. This probabilistic modeling approach was later generalized in [8] to model both jamming and DoS attacks as well as random transmission failures on multi-hop networked control systems. Despite the probabilistic nature of the model, the timing of attacks can still be arbitrary given that they satisfy certain conditions on the average attack duration.

Probabilistic approaches in networked control systems also include the SINR-based models, which can provide more detailed descriptions of failures on wireless channels. In [30, 31, 48], SINR-based models were utilized in a game-theoretic formulation of a state estimation problem under jamming attacks. More recently, in [7], we used an SINR-based wireless channel model to study a networked control problem under jamming attacks. The model there allows the likelihood of transmission failures to depend on jamming interference power. We showed that asymptotic stabilization can be achieved through networked control, if the jamming interference power stays bounded in average even if it can be very large at certain times. The problem setting in that work allows the interference power to depend on time but not the past and present values of the state. This setting facilitates the analysis; however, it also restricts the practical class of attacks that we can consider.

In this chapter, our goal is to explore a much wider class of attacks than the one that we considered previously in [7]. In particular, we want to investigate "state-dependent jamming attacks", where the attacker is more aware of the system and is allowed to select the jamming interference power at each time based on the past and the present state of the system. The results in [7] cannot deal with such state-dependent jamming, since some of the technical results there require certain independence assumptions, which are not satisfied when attackers can use the state information. However, we prove in the following sections that the almost-sure asymptotic stability result of [7] can still be recovered, and this is done without introducing any additional conservatism. As the first step of our stability analysis, we show that the difference between the total number of transmission failures and the sum of interference-power-dependent conditional transmission failure probabilities is a martingale. Then by using Azuma's inequality for martingales, we analyze the effects of state-dependent jamming interference on stability. We obtain sufficient stability conditions, which indicate that stability can be achieved when the attacker has energy constraints so that the average jamming interference power has a sufficiently small upper bound.

We illustrate our results through an optimization-based state-dependent attack strategy. The attacker's goal in this strategy is to maximize the sum of the second moments of predicted future state values. We consider the scenario where the attacker uses the current state information together with a *rolling-horizon* optimization approach. This approach is similar to the model predictive or the rolling-horizon control approaches: At each time, the attacker plans ahead by obtaining optimal jamming interference powers for a given finite time horizon but applies the first interference power. Since there are two outcomes depending on whether the interference causes transmission failure or not, this approach allows the attacker to update the state information to get better future state predictions. The optimization problem is then solved with the updated predictions in the following time steps again repetitively. The rolling-horizon attack strategy is demonstrated to be effective to identify damaging attacks for a given networked control system. In this regard, it fits into a stochastic version of the so-called *antagonistic control* framework discussed in [32]. We illustrate our approach through several simulations. In these simulations, we compare a number of different scenarios and observe that attackers with more computational resources can jam the networks more efficiently and cause bigger issues in the networked control system's operation.

We organize the remaining sections as follows. In Sect. 3.2, we discuss networked control system dynamics, SINR-based wireless channel model, and the attacker's capabilities. We explain an optimization-based rolling-horizon jamming attack strategy in Sect. 3.3. We then analyze the stability of the closed-loop networked control systems under jamming attacks in Sect. 3.4. Then in Sect. 3.5, we demonstrate the effects of rolling-horizon attack strategy through a numerical example. We finally provide some concluding remarks in Sect. 3.6. Some of the results in Sect. 3.4 were presented in our short conference report [9] without proofs. Here, in this work, we provide the complete proofs together with new discussions and examples.

Notation: Throughout this chapter, we denote nonnegative and positive integers by $\mathbb{N}_0$ and $\mathbb{N}$, respectively. Moreover, we denote the Euclidean norm by $\|\cdot\|$ and use $\lceil\cdot\rceil$ to represent the smallest integer that is larger than or equal to its real argument. We use the notations $\mathbb{E}[\cdot]$ and $\mathbb{P}[\cdot]$ to denote the expectation and probability measure on a given probability space $(\Omega, \mathscr{F}, \mathbb{P})$. In addition, the indicator of an event $F \in$

$\mathscr{F}$ is denoted by $\mathbb{1}[F] : \Omega \to \{0, 1\}$, which satisfies $\mathbb{1}[F](\omega) = 1$ for $\omega \in F$, and $\mathbb{1}[F](\omega) = 0$ for $\omega \notin F$. We omit $\omega \in \Omega$ in the notation of random variables in certain equations to simplify the presentation.

3.2 Networked Control Under State-Dependent Jamming

We consider the state-feedback wireless networked control problem of a discrete-time linear plant. In the networked control operation, at each time step t, the controller receives the state information $x(t)$, computes a control input $Kx(t)$, and attempts to transmit it on a wireless channel (Fig. 3.1). This channel is assumed to be subject to jamming attacks by an attacker who emits interference signals to cause transmission failures.

If the transmission at time t fails, then the control input at the plant side is set to 0; if the transmission is successful, then the transmitted control input is applied directly. Thus, the closed-loop system dynamics is given by

$$x(t+1) = Ax(t) + Bu(t), \quad x(0) = x_0, \quad t \in \mathbb{N}_0, \tag{3.1}$$

$$u(t) = (1 - l(t))Kx(t), \tag{3.2}$$

where $A \in \mathbb{R}^{n \times n}$ is the unstable system matrix, $B \in \mathbb{R}^{n \times m}$ is the input matrix, $x(t) \in \mathbb{R}^n$ is the state, $u(t) \in \mathbb{R}^m$ is the control input applied at the plant side, and $l(t) \in \{0, 1\}$ is the transmission failure indicator (with $l(t) = 1$ representing a failed transmission and $l(t) = 0$ representing a successful transmission). Moreover, the power of the jamming interference signal at time t is denoted by $v(t) \in [0, \infty)$.

The relation between the transmission failure indicator $l(t)$ and the jamming interference power $v(t)$ is described by

$$l(t) \triangleq \mathbb{1}[r(t) \le p(v(t))], \quad t \in \mathbb{N}_0, \tag{3.3}$$

where $p\colon [0, \infty) \to [0, 1]$ is a Borel-measurable and nondecreasing function, and moreover, $r(0), r(1), \ldots$ are independent random variables that are distributed uniformly in $[0, 1]$.

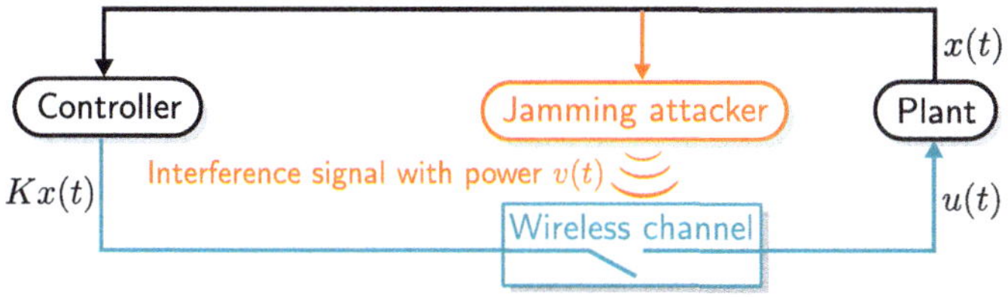

Fig. 3.1 Operation of networked control system under jamming attacks

The characterization in (3.3) describes the effect of the jamming interference power on the likelihood of a transmission failure. If $v(t)$ is large so that $p(v(t))$ is close to 1, then it becomes more likely that $r(t) \leq p(v(t))$, and hence by (3.3), a transmission failure ($l(t) = 1$) becomes more likely. Here, the use of the uniformly distributed random variables $r(\cdot)$ in the model allows us to specify detailed dependency structures for characterizing various jamming attack strategies such as time-dependent and state-dependent attacks.

We call $p(\cdot)$ in (3.3) the *conditional transmission failure probability function*, since (3.3) implies $\mathbb{P}[l(t) = 1|v(t)] = p(v(t))$ for every $t \in \mathbb{N}_0$.

3.2.1 Transmission Failure Probabilities and Their Relation to Signal-to-Interference-plus-Noise Ratio

The interference model in (3.3) allows us to describe the security and the reliability properties of different wireless channel models through different $p(\cdot)$ functions. For example, consider the additive white Gaussian noise channel with quadrature amplitude modulation scheme considered in the work [30]. The failures on this channel can be described by

$$p(\nu) = 2Q\Big(\sqrt{c\frac{\xi}{\nu+\sigma}}\Big), \quad \nu \in [0, \infty), \tag{3.4}$$

with $Q(y) \triangleq \frac{1}{\sqrt{2\pi}} \int_y^{\infty} e^{-\frac{s^2}{2}} \mathrm{d}s$, where $\xi \in (0, \infty)$ and $\sigma \in (0, \infty)$ are constants associated, respectively, with the power of the transmitted signal and the power of the channel noise, and moreover, $c \in (0, \infty)$ is a constant parameter of the communication protocol. The term $\frac{\xi}{\nu+\sigma}$ in (3.4) corresponds to Signal-to-Interference-plus-Noise Ratio (SINR), which is an indicator of the quality of a wireless channel [44].

SINR-based models have become indispensable in the characterization of wireless channels under malicious or nonmalicious and unintentional interference. In these models, the transmission failure probability is given as a function of SINR. The specific form of the function depends on the channel types and parameters such as the modulation scheme, size of the packet to be transmitted, and channel gains, which, in turn, depend on the distances between the transmitter, the receiver, and the interference sources [22, 44]. The framework developed in this paper can deal with those different channel types and parameters, as long as the corresponding (measurable and nondecreasing) function p is available for analysis.

We remark that in addition to SINR-based models, our setup allows us to also recover the scenarios where the actions of the attacker directly affect the failure. Specifically, in a setting with $p(v) = v$, $v \in [0, 1]$ and $p(v) = 1$, $v > 1$, the failure indicator satisfies $\mathbb{P}[l(t) = v(t)] = 1$ for attacks with $v(t) \in \{0, 1\}$. This setting is useful for modeling scenarios where the attacker can turn off the jamming or attack with a very large interference power that causes failures almost surely.

Notice that if the interference power is fixed at a constant level ϑ^* and does not change as time progresses (i.e., $v(t) = \vartheta^*$, $t \in \mathbb{N}_0$), then the model that we describe with (3.3) reduces to Bernoulli-type packet losses with packet loss probability $p(\vartheta^*)$. The Bernoulli model has been studied extensively in the past (see, e.g., [25, 26, 40]). We deviate from the Bernoulli setting. As we explain in the following subsection, our goal is to investigate the scenarios where the attacker can manipulate the interference power $v(t)$ in a strategic way.

3.2.2 Attacker's Capabilities and Constraints in State-Dependent Jamming

We would like to investigate the effects of *state-dependent* jamming on the networked control system (3.1), (3.2). To this end, we consider a general setup, where we describe the attacker's capabilities by utilizing the filtration $\{\mathscr{F}_t\}_{t\in\mathbb{N}_0}$, where for each $t \in \mathbb{N}_0$, $\mathscr{F}_t$ is the σ-algebra generated by the random variables $r(0), r(1), \ldots, r(t)$ and $v(0), v(1), \ldots, v(t)$ appearing in the attack model (3.3). We assume that for every $t \in \mathbb{N}_0$, the random variables $r(t)$ and $v(t)$ are mutually independent. Moreover, for every $t \in \mathbb{N}$, $r(t)$ is assumed to be independent of $\mathscr{F}_{t-1}$.

Under the assumptions mentioned above, the interference power $v(t)$ is allowed to depend on $r(0), r(1), \ldots, r(t-1)$ as well as $v(0), v(1), \ldots, v(t-1)$. A direct consequence is that we can consider jamming attacks that depend on previous packet transmission failure indicators as well as *state-dependent jamming*, where $v(t)$ depends on $x(t)$, which is a function of $r(0), r(1), \ldots, r(t-1)$ and $v(0), v(1), \ldots, v(t-1)$, namely

$$\begin{aligned} x(t) = {} & (A + (1 - \mathbb{1}[r(t-1) \leq p(v(t-1))])BK) \\ & \cdot (A + (1 - \mathbb{1}[r(t-2) \leq p(v(t-2))])BK) \\ & \cdots (A + (1 - \mathbb{1}[r(0) \leq p(v(0))])BK)x_0. \end{aligned} \tag{3.5}$$

Here x_0 is considered as a fixed nonrandom vector known to the attacker. Notice that an attacker who has access to the state information $x_0, x(1), x(2), \ldots$ can utilize this information at each time to strategically pick the power level of the interference signal.

Generation and emission of interference signals are energy-consuming actions and costly to the attacker [39, 47]. Thus, the attacker cannot use very large interference powers at all times. Here, we take this observation into account and consider the scenarios where the *average interference power of the attacker* is bounded. In particular, following an approach similar to our previous work [7], we assume that the interference power process $\{v(t) \in [0, \infty)\}_{t\in\mathbb{N}_0}$ satisfies the following assumption.

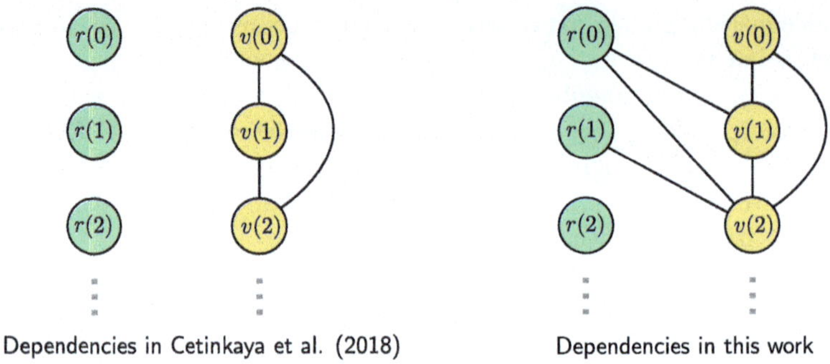

Fig. 3.2 Comparison of Cetinkaya et al. [7] and this work with respect to the dependency graphs of random variables $r(0), r(1), r(2), v(0), v(1), v(2)$

Assumption 3.1 There exist scalars $\kappa, \overline{v} \geq 0$ such that

$$\mathbb{P}\left[\sum_{i=0}^{t-1} v(i) \leq \kappa + \overline{v}t\right] = 1, \quad t \in \mathbb{N}. \tag{3.6}$$

The nonnegative scalar $\overline{v}$ in this assumption is an asymptotic upper bound on the average interference power. Specifically, we have $\limsup_{k\to\infty} \frac{1}{k}\sum_{t=0}^{k-1} v(t) \leq \overline{v}$, almost surely. The scalar κ has a different role and it allows modeling the initial behavior of the attacks. Notice that when $\kappa = 0$, (3.6) implies $v(t) \leq (t+1)\overline{v}$. Scenarios where $v(t) > (t+1)\overline{v}$ for some $t \in \mathbb{N}_0$ can be modeled by setting $\kappa > 0$.

With Assumption 3.1, we can characterize the average strength of an attack without the need of specifying particular actions of the attacker at each time. This characterization is similar to those in [15] and our previous work [6], where the strengths of attacks were described through the average duration of the attacks and the average amount of failures they cause. The approach in this work is different, since we use the interference power levels to identify the strength of an attack. Moreover, here we allow attacks to happen at each time as long as (3.6) is satisfied.

3.2.3 A Comparison of the Setups for State-Dependent and Time-Dependent Jamming

In our previous work [7], we also considered jamming attacks with time-varying interference powers. However, the setup that we consider now in this work is strictly more general. Specifically, in [7], $\{r(t) \in [0, 1]\}_{t\in\mathbb{N}_0}$ and $\{v(t) \in [0, \infty)\}_{t\in\mathbb{N}_0}$ were assumed to be completely independent processes, and as a result, state-dependent jamming attacks were excluded from the analysis.

A comparison of [7] and this work is given in Fig. 3.2 in terms of the dependency graphs of $r(t)$ and $v(t)$ for the first three time steps ($t \in \{0, 1, 2\}$). A *dependency graph* is an undirected graph, where the nodes represent random variables (see [2]). If there is no edge between two nodes, then the random variables associated with those nodes must be mutually independent. On the other hand, an edge indicates possible dependence. Notice that mutual independence and dependence are commutative relations, but they are not transitive. In this work, the interference power $v(2)$ is allowed to depend on $r(0), r(1), v(0), v(1)$. This is indicated on the right-hand side of Fig. 3.2 through the edges connected to the node $v(2)$. Hence by (3.5), $v(2)$ is allowed to be selected by the attacker based on the state information $x(2)$. However, an interference power in [7] can only depend on the previous interference powers, but *not* on the state (since $r(\cdot)$ and $v(\cdot)$ on the left-hand side of Fig. 3.2 are not connected).

The stability results in [7] indicate that if the upper bound $\overline{v}$ of the average interference powers is sufficiently small, then the networked control system is guaranteed to be stable. In the following section, we show that the sufficient stability conditions presented in [7] are also sufficient to ensure stability under state-dependent jamming attacks considered in this work. However, the analysis here differs from that in [7]. Specifically, the stability was established in [7] by utilizing

$$\mathbb{E}\left[\prod_{i=t_1}^{t_2-1}(\alpha_1 l(i) + \alpha_0)\right] = \mathbb{E}\left[\prod_{i=t_1}^{t_2-1}(\alpha_1 p(v(i)) + \alpha_0)\right], \tag{3.7}$$

which holds in the setting there for every $\alpha_0, \alpha_1 \in \mathbb{R}$, $t_1 \in \mathbb{N}_0$, $t_2 \in \{t_1 + 1, t_1 + 2, \ldots\}$. This equality establishes a special relation between the transmission failure indicator $l(\cdot)$ and the attacker's interference power $v(\cdot)$. We remark that (3.7) holds and can be used in the analysis when $\{r(t) \in [0, 1]\}_{t\in\mathbb{N}_0}$ and $\{v(t) \in [0, \infty)\}_{t\in\mathbb{N}_0}$ are independent processes. This is, for example, the case when $v(t)$ is a deterministic function of time. However, in our current setup, $\{r(t) \in [0, 1]\}_{t\in\mathbb{N}_0}$ and $\{v(t) \in [0, \infty)\}_{t\in\mathbb{N}_0}$ are not necessarily independent as mentioned before. As a result, (3.7) does not always hold.

As an example, consider the simple case with a strictly increasing transmission failure probability function p with $p(0) \in (0, 1)$ and an attack strategy given by $v(0) = 0$ and

$$v(t) = \begin{cases} \overline{v}, & l(t-1) = 0, \\ 0, & l(t-1) = 1, \end{cases} \quad t \in \mathbb{N}, \tag{3.8}$$

with $\overline{v} > 0$. This attack strategy turns the jamming on and off based on the packet transmission status at the previous time instant. If the jammer has knowledge of the dynamical system and can access to the state information, then $l(t-1)$ can be identified from the state values $x(t)$ and $x(t-1)$ whenever $BKx(t-1) \neq 0$. Particularly, the noise-free dynamics (3.1) and (3.2) imply that $l(t-1) = 1$ if $x(t) =$

$Ax(t-1)$, and $l(t-1)=0$, otherwise. In this setting, by the law of total expectation (see Theorem A.3 of [38]), we have

$$\begin{aligned}\mathbb{E}[l(0)l(1)] &= \mathbb{E}[\mathbb{E}[l(0)l(1) \mid l(0)]] \\ &= \mathbb{E}[l(0)\mathbb{E}[l(1) \mid l(0)]] \\ &= \mathbb{E}[l(0)\mathbb{E}[\mathbb{1}[r(1) \leq p(v(1))] \mid l(0)]] \\ &= \mathbb{E}[l(0)\mathbb{E}[\mathbb{1}[r(1) \leq p((1-l(0))\overline{v})] \mid l(0)]] \\ &= \mathbb{E}[l(0)p((1-l(0))\overline{v})] = p^2(0). \end{aligned} \tag{3.9}$$

On the other hand,

$$\begin{aligned}\mathbb{E}[p(v(0))p(v(1))] &= p(0)\mathbb{E}[p(v(1))] \\ &= p(0)\mathbb{E}[p((1-l(0))\overline{v})] \\ &= p(0)\left(p(0)p(0)+p(\overline{v})(1-p(0))\right) > p^2(0), \end{aligned} \tag{3.10}$$

since $p(\overline{v}) > p(0) > 0$. From (3.9) and (3.10), we observe that (3.7) does not hold for $\alpha_1 = 1, \alpha_2 = 0, t_1 = 0, t_2 = 2$, under the attack strategy (3.8).

As in the example above, (3.7) cannot be used in the analysis for many other attack scenarios. In the next section, we discuss such a scenario, where an adversary follows an optimization approach to form a state-dependent jamming strategy.

3.3 Rolling-Horizon Jamming Attacks: A State-Dependent Optimization-Based Jamming Strategy

In this section, we introduce an *optimization-based* strategy for the attacker for maximizing the effect of *jamming* when he/she is aware of the system state. Consider the scenario where the attacker has access to the state information at each time t. The attacker may be able to observe the state directly or it may be the case that the attacker knows the initial state x_0 and reconstructs the state at each time by using the control system dynamics and packet failure states for all earlier time steps $0, \ldots, t-1$. For this setup, we investigate the attack strategy, where the attacker decides the interference power at time t, by maximizing a utility function associated with predicted future state values.

In particular, the attacker uses the value x_t of the state at time t to decide $v(t)$ according to

$$v(t) = v_t^*, \tag{3.11}$$

where $(v_t^*, v_{t+1}^*, \ldots, v_{t+T_A-1}^*) \in [0, \infty)^{T_A}$ is a solution to the optimization problem

$$\begin{aligned}
&\underset{(v_t,\ldots,v_{t+T_A-1})}{\text{maximize}} \quad U(x_t, v_t, \ldots, v_{t+T_A-1}) \\
&\text{subject to} \quad v_{\text{sum}}(t) + \sum_{i=t}^{t+j-1} v_i \le \kappa + \overline{v}(t+j), \\
&\qquad\qquad\quad v_{t+j-1} \ge 0, \quad j \in \{1, \ldots, T_A\},
\end{aligned} \tag{3.12}$$

where $\kappa \ge 0$, $\overline{v} \ge 0$, and $T_A \in \mathbb{N}$; moreover, $U(x_t, v_t, \ldots, v_{t+T_A-1})$ denotes the attacker's utility function, and $v_{\text{sum}}(t)$ denotes the sum of interference powers already used by the attacker for the first t time steps, i.e.

$$v_{\text{sum}}(t) \triangleq \begin{cases} 0, & t = 0, \\ \sum_{i=0}^{t-1} v(i), & t \in \mathbb{N}. \end{cases} \tag{3.13}$$

Here the attacker's goal is to maximize a utility function while keeping the average interference power below a certain value. In particular, under this attack strategy, interference powers always satisfy Assumption 3.1 with $\kappa \ge 0$ and $\overline{v} \ge 0$.

The scalar $T_A \in \mathbb{N}$ is the length of the horizon in the optimization problem (3.12), and it can be large if the attacker has sufficient amount of computational resources. The optimization is done in the rolling-horizon fashion, where at time t, the attacker plans ahead and obtains T_A number of interference powers $(v_t^*, v_{t+1}^*, \ldots, v_{t+T_A-1}^*)$ but applies the first interference power (i.e., $v(t) = v_t^*$). In the next time step, the state information is updated and the optimization problem is solved again as the updated information allows better prediction of the future states.

The specific form of the utility function $U(\cdot)$ that we consider is described as follows. Let Q_h be the set of binary sequences of length h, that is

$$Q_h \triangleq \{(q_1, q_2, \ldots, q_h) \colon q_j \in \{0, 1\}, j \in \{1, \ldots, h\}\}. \tag{3.14}$$

Furthermore, let

$$\hat{A}(s) \triangleq sA + (1-s)(A + BK), \quad s \in \{0, 1\}, \tag{3.15}$$

$$\hat{F}_h(q) \triangleq \hat{A}(q_h) \cdots \hat{A}(q_1), \quad q \in Q_h, \tag{3.16}$$

and for each $h \in \{1, \ldots, T_A\}$, define

$$\begin{aligned}
&U_h(x_t, v_t, \ldots, v_{t+T_A-1}) \\
&\triangleq \sum_{q \in Q_h} \|\hat{F}_h(q)x_t\|^2 \Big(\prod_{i=1}^{h} \big(q_i p(v_{t+i-1}) + (1-q_i)(1 - p(v_{t+i-1}))\big) \Big).
\end{aligned} \tag{3.17}$$

The utility function is then given by

$$U(x_t, v_t, \ldots, v_{t+T_A-1}) \triangleq \sum_{h=1}^{T_A} U_h(x_t, v_t, \ldots, v_{t+T_A-1}). \tag{3.18}$$

The term $\prod_{i=1}^{h} \left(q_i p(v_{t+i-1}) + (1-q_i)(1-p(v_{t+i-1}))\right)$ in (3.17) represents the probability that the transmission failure indicator $l(\cdot)$ takes the particular binary values $q_1, q_2, \ldots, q_h$ at time steps $t, t+1, \ldots t+h-1$, given that the attacker jams the channel with power levels $v_t, v_{t+1}, \ldots, v_{t+h-1}$ at those times. Notice that when the transmission failure indicator takes those values, the state value given as x_t at time t, becomes $\hat{F}_h(q)x_t$ at time $t+h$.

Intuitively, $U_h(\cdot)$ in (3.17) represents the conditional expectation of the squared norm (conditional second moment) of the state at time $t+h$, given the information of jamming power levels for times $t, t+1, \ldots t+h-1$ and the state at time t. Consequently, as the following proposition indicates, the utility function $U(\cdot)$ in (3.18) corresponds to the sum of future conditional second moments of the state, given the state value x_t and the particular interference powers $v_t, \ldots, v_{t+T_A-1}$.

Proposition 3.1 *Let $G \in \mathscr{F}_t$ be the event given by*

$$G \triangleq \{\omega \in \Omega\colon x(t)=x_t, v(t)=v_t, v(t+1) = v_{t+1}, \ldots, v(t+T_A-1) = v_{t+T_A-1}\},$$

where $x_t \in \mathbb{R}^n$, and $v_i \in [0, \infty)$, $i \in \{t, \ldots, t+T_A-1\}$. Assume $\mathbb{1}[G]$ is independent of $r(t), r(t+1), \ldots, r(t+T_A-1)$, and moreover, $\mathbb{P}[G] = 1$. Then we have

$$\mathbb{E}\left[\sum_{h=1}^{T_A} \|x(t+h)\|^2 \,\middle|\, G\right] = U(x_t, v_t, \ldots, v_{t+T_A-1}). \tag{3.19}$$

Proof It follows from (3.15) that

$$\begin{aligned} \|x(t+h)\|^2 &= \|\hat{A}(l(t+h-1)\cdots\hat{A}(l(t))x(t)\|^2 \\ &= \sum_{q\in Q_h} \|\hat{A}(q_h)\cdots\hat{A}(q_1)x(t)\|^2 \left(\prod_{i=1}^{h} \mathbb{1}[l(t+i-1)=q_i]\right). \end{aligned}$$

Hence, by using (3.16) and then interchanging the order of summation and expectation (through Fubini's theorem [5])

$$\mathbb{E}[\|x(t+h)\|^2\mathbb{1}[G]] = \sum_{q\in Q_h} \mathbb{E}\left[\|\hat{F}_h(q)x(t)\|^2 \left(\prod_{i=1}^{h} \mathbb{1}[l(t+i-1)=q_i]\right)\mathbb{1}[G]\right]. \tag{3.20}$$

By (3.3), the term $\mathbb{1}[l(t+i-1)=q_i]$ can be represented as

$$\mathbb{1}[l(t+i-1)=q_i] = q_i\mathbb{1}[r(t+i-1)\le p(v(t+i-1)] \\ +(1-q_i)(1-\mathbb{1}[r(t+i-1)\le p(v(t+i-1)]). \quad (3.21)$$

Moreover, $\|\hat{F}_h(q)x(t)\|^2\mathbb{1}[G] = \|\hat{F}_h(q)x_t\|^2\mathbb{1}[G]$ and

$$\mathbb{1}\left[r(t+i-1)\le p(v(t+i-1)\right]\mathbb{1}[G] = \mathbb{1}[r(t+i-1)\le p(v_{t+i-1})]\mathbb{1}[G], \quad (3.22)$$

almost surely. To simplify the notation, let

$$\xi_i \triangleq q_i\mathbb{1}[r(t+i-1)\le p(v_{t+i-1})] + (1-q_i)(1-\mathbb{1}[r(t+i-1)\le p(v_{t+i-1})]),$$

for $i\in\{1,\ldots,h\}$. It follows from (3.20)–(3.22) that

$$\begin{aligned}\mathbb{E}\left[\|x(t+h)\|^2\mathbb{1}[G]\right] &= \sum_{q\in Q_h}\mathbb{E}\left[\|\hat{F}_h(q)x_t\|^2\left(\prod_{i=1}^{h}\xi_i\right)\mathbb{1}[G]\right] \\ &= \sum_{q\in Q_h}\|\hat{F}_h(q)x_t\|^2\mathbb{E}\left[\left(\prod_{i=1}^{h}\xi_i\right)\mathbb{1}[G]\right]. \quad (3.23)\end{aligned}$$

Since $\mathbb{1}[G]$ is independent of $r(t), r(t+1),\ldots,r(t+T_{\mathrm{A}}-1)$, it is independent of ξ_i, $i\in\{1,\ldots,h\}$, as well. Hence, $\mathbb{E}\Big[\big(\prod_{i=1}^{h}\xi_i\big)\mathbb{1}[G]\Big] = \mathbb{E}\big[\big(\prod_{i=1}^{h}\xi_i\big)\big]\mathbb{E}\big[\mathbb{1}[G]\big]$. Since $r(t), r(t+1),\ldots,r(t+T_{\mathrm{A}}-1)$ are mutually independent, it follows that ξ_i, $i\in\{1,\ldots,h\}$, are also mutually independent. As a result, $\mathbb{E}\big[\big(\prod_{i=1}^{h}\xi_i\big)\big] = \prod_{i=1}^{h}\mathbb{E}[\xi_i]$. By using these identities and noting that $\mathbb{E}\big[\mathbb{1}[G]\big]=\mathbb{P}[G]=1$, we obtain from (3.23)

$$\mathbb{E}[\|x(t+h)\|^2\mathbb{1}[G]] = \sum_{q\in Q_h}\|\hat{F}_h(q)x_t\|^2\left(\prod_{i=1}^{h}\mathbb{E}[\xi_i]\right).$$

Now, by $\mathbb{E}[\xi_i] = q_i p(v_{t+i-1}) + (1-q_i)(1-p(v_{t+i-1}))$

$$\mathbb{E}[\|x(t+h)\|^2\mathbb{1}[G]] = U_h(x_t, v_t,\ldots,v_{t+T_{\mathrm{A}}-1}). \quad (3.24)$$

Since $\mathbb{P}[G]=1$, by using Fubini's theorem again, we get

$$\begin{aligned}\mathbb{E}\left[\sum_{h=1}^{T_{\mathrm{A}}}\|x(t+h)\|^2 \;\middle|\; G\right] &= \mathbb{E}\left[\sum_{h=1}^{T_{\mathrm{A}}}\|x(t+h)\|^2\mathbb{1}[G]\right] \\ &= \sum_{h=1}^{T_{\mathrm{A}}}\mathbb{E}\left[\|x(t+h)\|^2\mathbb{1}[G]\right]. \quad (3.25)\end{aligned}$$

Finally by using (3.18), (3.24), and (3.25), we obtain (3.19). □

Notice that $U(\cdot)$ given in (3.18) is not necessarily a concave function of the variables $v_t, \ldots, v_{t+T_A-1}$, which makes (3.12) a nonconvex optimization problem. However, if p is a continuous function (e.g., (3.4)), then $U(\cdot)$ is a continuous function. It follows from the compactness of the constraint set for $v_t, \ldots, v_{t+T_A-1}$ and the extreme value theorem (see Theorem 4.16 of [41]) that the optimization problem (3.14) possesses optimal solutions. In practice, we can use nonconvex optimization techniques to numerically find $v_t, \ldots, v_{t+T_A-1}$ that result in large values of $U(\cdot)$. Even though such $v_t, \ldots, v_{t+T_A-1}$ may not be optimal, they can allow us to identify potentially dangerous attacks.

It is important to be able to assess the cybersecurity of networked control systems under state-dependent jamming attacks including the optimization-based rolling-horizon attacks discussed in this section. In this chapter we show that closed-loop stability can be guaranteed even under state-dependent attacks, as long as the jamming interference power level is constrained as in Assumption 3.1 and the average interference power bound is sufficiently small. This is elaborated further in the next section, where we introduce new techniques for stability analysis under state-dependent jamming attacks with interference power constraints.

3.4 Analysis of Networked Stabilization Under State-Dependent Jamming Attacks

In this section, we provide an analysis of the networked control system (3.1), (3.2) under state-dependent jamming. Our goal is to obtain conditions under which the state converges to the origin almost surely (i.e., $\mathbb{P}[\lim_{t\to\infty} \|x(t)\| = 0] = 1$). To this end, we consider the *almost-sure asymptotic stability* notion [6, 46].

Definition 3.1 The zero solution $x(t) \equiv 0$ of the discrete-time stochastic system (3.1), (3.2) is *almost surely stable* if, for every $\varepsilon > 0$ and $\bar{p} > 0$, there exists a positive scalar $\delta = \delta(\varepsilon, \bar{p}) > 0$ such that if $\|x(0)\| < \delta$, then

$$\mathbb{P}\left[\sup_{t\in\mathbb{N}_0} \|x(t)\| > \varepsilon\right] < \bar{p}. \tag{3.26}$$

Furthermore, the zero solution $x(t) \equiv 0$ is *asymptotically stable almost surely* if it is almost surely stable and

$$\mathbb{P}\left[\lim_{t\to\infty} \|x(t)\| = 0\right] = 1. \tag{3.27}$$

For almost-sure asymptotic stability analysis, the function $p\colon [0, \infty) \to [0, 1]$ in (3.3) plays an essential role. In particular, if $p(\cdot)$ is a concave function, then

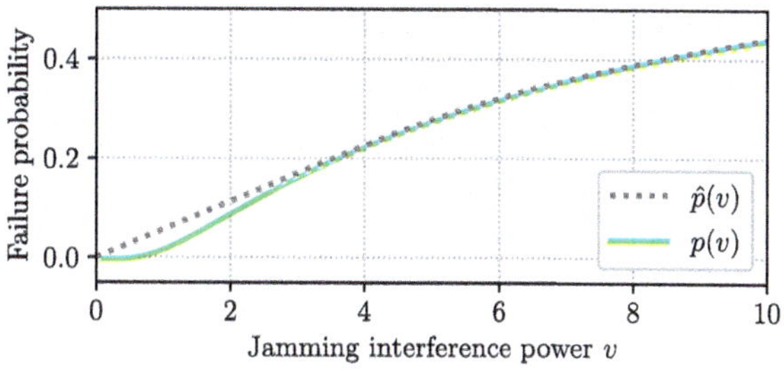

Fig. 3.3 Comparison of conditional transmission failure probability function p and its concave upper bound $\hat{p}$

$$\frac{1}{t}\sum_{i=0}^{t-1} p(v(i)) \le p\left(\frac{1}{t}\sum_{i=0}^{t-1} v(i)\right), \tag{3.28}$$

holds. This relation indicates that the average of the conditional transmission failure probabilities at different times is bounded by the conditional transmission failure probability associated with the average interference power. However, depending on the considered wireless channel, $p(\cdot)$ may not always be a concave function. In such cases, (3.28) does not necessarily hold, but a concave function that upper-bounds $p(\cdot)$ can be used in the analysis. Let $\hat{p}\colon [0,\infty) \to [0,1]$ be a continuous, nondecreasing, and concave function such that

$$\hat{p}(v) \ge p(v), \quad v \in [0,\infty). \tag{3.29}$$

Note that $\hat{p}(\cdot)$ satisfying (3.29) always exists.

As our main result below indicates, even under state-dependent jamming attacks, the stability can be assessed by checking certain conditions that involve the scalar $\hat{p}(\overline{v})$ with $\overline{v} \ge 0$ denoting the bound of the average interference power in Assumption 3.1. In particular, stability is ensured if $\hat{p}(\overline{v})$ is sufficiently small. In order not to introduce conservatism, it is ideal to obtain $\hat{p}$ as a *tight* upper bound of p. The works [7, 37] introduce several approaches to obtain $\hat{p}$. For the example p given in (3.4) with $c = 1$, $\xi = 6$, and $\sigma = 0.1$, a tight concave upper bound $\hat{p}$ is shown in Fig. 3.3.

We are now ready to state our main result concerning stability.

Theorem 3.1 *Consider the networked control system (3.1), (3.2). Suppose that the jamming interference power process* $\{v(t) \in [0,\infty)\}_{t\in\mathbb{N}_0}$ *satisfies Assumption 3.1. If there exist a positive-definite matrix* $P \in \mathbb{R}^{n\times n}$ *and scalars* $\beta \in (0,1)$, $\varphi \in [1,\infty)$ *such that*

$$\beta P - (A+BK)^{\mathrm{T}} P (A+BK) \ge 0, \tag{3.30}$$

$$\varphi P - A^{\mathrm{T}} P A \ge 0, \tag{3.31}$$

$$(1-\hat{p}(\overline{v}))\ln\beta + \hat{p}(\overline{v})\ln\varphi < 0, \tag{3.32}$$

then the zero solution $x(t) \equiv 0$ *of the closed-loop system (3.1), (3.2) is asymptotically stable almost surely.*

Theorem 3.1 provides sufficient conditions for the stability of the system (3.1), (3.2). In conditions (3.30) and (3.31), we use the scalars β and φ, respectively, to characterize the stability of the closed-loop dynamics and the instability of the open-loop dynamics. The condition (3.32) contains β and φ together with $\hat{p}(\overline{v})$ representing the effect of the jamming attacks. When the average interference power $\overline{v}$ is sufficiently small, then $\hat{p}(\overline{v})$ takes a small value and (3.32) is satisfied, implying stability.

The proof of Theorem 3.1 utilizes the stochastic Lyapunov-based argument in [6], which requires us to obtain a long-run upper bound of average number $\frac{1}{t}\sum_{i=0}^{t-1} l(i)$ of transmission failures. In what follows, our goal is to show that this average number is in fact upper-bounded by $\hat{p}(\overline{v})$ in the long run. We derive this upper bound by obtaining (1) a probabilistic relation between $\hat{p}(\overline{v})t$ and $\sum_{i=0}^{t-1} \hat{p}(v(i))$, and (2) a probabilistic relation between $\sum_{i=0}^{t-1} \hat{p}(v(i))$ and $\sum_{i=0}^{t-1} l(i)$. These two relations are then combined in the proof of Theorem 3.1 to show $\limsup_{t\to\infty} \frac{1}{t}\sum_{i=0}^{t-1} l(i) \le \hat{p}(\overline{v})$, almost surely. The proofs of these relations rely on properties of the concave function $\hat{p}$ and a concentration bound for martingales. Our results indicate that even though the attack strategy is uncertain and the failure probabilities at each time are unknown, the average interference power bound $\overline{v}$ can be used to assess stability.

The following result characterizes certain properties of the jamming interference power process $\{v(t) \in [0,\infty)\}_{t\in\mathbb{N}_0}$ under Assumption 3.1 and establishes a relation between $\hat{p}(\overline{v})t$ and $\sum_{i=0}^{t-1} \hat{p}(v(i))$.

Lemma 3.1 *Suppose that the jamming interference power process* $\{v(t) \in [0,\infty)\}_{t\in\mathbb{N}_0}$ *satisfies Assumption 3.1 with* $\kappa \ge 0$ *and* $\overline{v} \ge 0$. *Then, we have*

$$\mathbb{P}\left[\sum_{i=0}^{t-1} \hat{p}(v(i)) \le \hat{p}(\frac{\kappa}{t} + \overline{v})t\right] = 1, \quad t \in \mathbb{N}. \tag{3.33}$$

Moreover, for every $\tilde{\varepsilon} > 0$, *there exists* $\tilde{\tau} \in \mathbb{N}$ *such that*

$$\mathbb{P}\left[\sum_{i=0}^{t-1} \hat{p}(v(i)) - \hat{p}(\overline{v})t > \tilde{\varepsilon}t\right] = 0, \quad t \ge \tilde{\tau}. \tag{3.34}$$

Proof Assumption 3.1 implies that

$$\mathbb{P}\left[\frac{1}{t}\sum_{i=0}^{t-1} v(i) \le \frac{\kappa}{t} + \overline{v}\right] = 1, \quad t \in \mathbb{N}. \tag{3.35}$$

Moreover, since $\hat{p}$ is a nondecreasing function, we have

$$\left\{\omega \in \Omega : \hat{p}\left(\frac{1}{t}\sum_{i=0}^{t-1} v(i)\right) \le \hat{p}\left(\frac{\kappa}{t} + \overline{v}\right)\right\} \supseteq \left\{\omega \in \Omega : \frac{1}{t}\sum_{i=0}^{t-1} v(i) \le \frac{\kappa}{t} + \overline{v}\right\}.$$

Hence, it follows from (3.35) that

$$\mathbb{P}\left[\hat{p}\left(\frac{1}{t}\sum_{i=0}^{t-1} v(i)\right) \leq \hat{p}\left(\frac{\kappa}{t}+\overline{v}\right)\right] = 1, \quad t \in \mathbb{N}. \tag{3.36}$$

Since $\hat{p}$ is a concave function, we have $\frac{1}{t}\sum_{i=0}^{t-1}\hat{p}(v(i)) \leq \hat{p}\left(\frac{1}{t}\sum_{i=0}^{t-1}v(i)\right)$, and thus, $\mathbb{P}\left[\frac{1}{t}\sum_{i=0}^{t-1}\hat{p}(v(i)) \leq \hat{p}\left(\frac{\kappa}{t}+\overline{v}\right)\right] \geq \mathbb{P}\left[\hat{p}(\frac{1}{t}\sum_{i=0}^{t-1}v(i)) \leq \hat{p}\left(\frac{\kappa}{t}+\overline{v}\right)\right]$. By using this inequality together with (3.36), we obtain (3.33).

Next, we show (3.34). For the case where $\kappa = 0$, (3.34) with $\tilde{\tau} = 1$ follows directly from (3.33). Now, consider the case where $\kappa > 0$. Since $\hat{p}(\cdot)$ is continuous and nondecreasing, there exists $\tilde{\delta} \in (0, \infty)$ (that depends on $\tilde{\varepsilon}$) such that $\hat{p}(\tilde{\delta}+\overline{v}) - \hat{p}(\overline{v}) < \tilde{\varepsilon}$. Since $\hat{p}$ is nondecreasing, with $\tilde{\tau} \triangleq \lceil \kappa/\tilde{\delta} \rceil$, we obtain $\hat{p}(\kappa/\tilde{\tau}+\overline{v}) - \hat{p}(\overline{v}) < \tilde{\varepsilon}$, and as a result

$$\hat{p}(\kappa/t+\overline{v}) - \hat{p}(\overline{v}) < \tilde{\varepsilon}, \quad t \geq \tilde{\tau}. \tag{3.37}$$

It then follows from (3.37) that for $t \geq \tilde{\tau}$

$$\mathbb{P}\left[\sum_{i=0}^{t-1}\hat{p}(v(i)) \leq \hat{p}\left(\frac{\kappa}{t}+\overline{v}\right)t\right] \leq \mathbb{P}\left[\sum_{i=0}^{t-1}\hat{p}(v(i)) \leq \hat{p}(\overline{v})t + \tilde{\varepsilon}t\right].$$

Using this inequality with (3.33), we obtain $\mathbb{P}[\sum_{i=0}^{t-1}\hat{p}(v(i)) \leq \hat{p}(\overline{v})t + \tilde{\varepsilon}t] = 1, t \geq \tilde{\tau}$, which implies (3.34). □

The equality (3.34) in Lemma 3.1 shows that for sufficiently large values of t, the scalar $\hat{p}(\overline{v})t$ is an accurate approximation of the sum $\sum_{i=0}^{t-1}\hat{p}(v(i))$. This indicates that the effect of the attack parameter κ diminishes and the long-term effects of state-dependent attacks can be represented through the upper bound $\overline{v}$ of the average interference powers.

Our next goal is to establish certain relations between the transmission failure indicator $l(\cdot)$ and the attacker's interference power $v(\cdot)$. To this end, we define the process $\{M(t) \in \mathbb{R}\}_{t\in\mathbb{N}_0}$ by

$$M(t) \triangleq \sum_{i=0}^{t} l(i) - \sum_{i=0}^{t} p(v(i)), \quad t \in \mathbb{N}_0. \tag{3.38}$$

Lemma 3.2 *The process $\{M(t) \in \mathbb{R}\}_{t\in\mathbb{N}_0}$ defined in (3.38) is a martingale with respect to the filtration $\{\mathscr{F}_t\}_{t\in\mathbb{N}_0}$.*

Proof The random variable $M(t)$ is $\mathscr{F}_t$-measurable, since the random variables $l(0), \ldots, l(t)$ and $v(0), \ldots, v(t)$ are. Next, we need to show $\mathbb{E}[M(t)|\mathscr{F}_{t-1}] = M(t-1)$, $t \in \mathbb{N}$. First, by (3.3)

$$\mathbb{E}[l(t) - p(v(t)) \mid \mathscr{F}_{t-1}] = \mathbb{E}[\mathbb{1}[r(t) \leq p(v(t))] - p(v(t)) \mid \mathscr{F}_{t-1}]. \tag{3.39}$$

Now, let $\mathscr{H}_t \triangleq \sigma(v(t), \mathscr{F}_{t-1})$ denote the σ-algebra generated by $v(t)$ and $\mathscr{F}_{t-1}$. Since p is a measurable function and $v(t)$ is $\mathscr{H}_t$-measurable, it follows that $p(v(t))$ is also $\mathscr{H}_t$-measurable. Furthermore, since $\mathscr{F}_{t-1} \subseteq \mathscr{H}_t$, by (3.39) and the law of total expectation (see Theorem A.3 of [38]),

$$\begin{aligned}\mathbb{E}[l(t) - p(v(t)) \mid \mathscr{F}_{t-1}] &= \mathbb{E}[\mathbb{E}[\mathbb{1}[r(t) \leq p(v(t))] - p(v(t)) \mid \mathscr{H}_t] \mid \mathscr{F}_{t-1}] \\ &= \mathbb{E}[\mathbb{E}[\mathbb{1}[r(t) \leq p(v(t))] \mid \mathscr{H}_t] - p(v(t)) \mid \mathscr{F}_{t-1}],\end{aligned} \tag{3.40}$$

almost surely. Our next goal is to show $\mathbb{E}[\mathbb{1}[r(t) \leq p(v(t))] | \mathscr{H}_t] = p(v(t))$. We use a set construction technique similar to the one utilized in Sect. 1.4 of [18] for obtaining inequalities that involve integrals with respect to measures. Let $\varepsilon \in (0, 1]$. Moreover, for a given $\varepsilon \in (0, 1]$, let $N(\varepsilon) \triangleq \lceil \frac{1}{\varepsilon} \rceil$. For $k \in \{1, \ldots, N(\varepsilon) + 1\}$, we define the events

$$E_k \triangleq \left\{ \omega \in \Omega : \, p(v(t)) \in \left[\frac{k-1}{N(\varepsilon)}, \frac{k}{N(\varepsilon)} \right) \right\}. \tag{3.41}$$

Here, we have $E_k \in \mathscr{H}_t$, since p is a measurable function and $v(t)$ is $\mathscr{H}_t$-measurable. Furthermore, $E_i \cap E_j = \emptyset$ for $i \neq j$, and moreover, $\cup_{k=1}^{N(\varepsilon)+1} E_k = \Omega$, since $p(v) \in [0, 1]$ for all $v \geq 0$. In addition

$$\sum_{k=1}^{N(\varepsilon)+1} \frac{k-1}{N(\varepsilon)} \mathbb{1}[E_k] \leq p(v(t)) \leq \sum_{k=1}^{N(\varepsilon)+1} \frac{k}{N(\varepsilon)} \mathbb{1}[E_k]. \tag{3.42}$$

Now, by using the properties of E_k

$$\begin{aligned}\mathbb{E}[\mathbb{1}[r(t) \leq p(v(t))] \mid \mathscr{H}_t] &= \mathbb{P}[r(t) \leq p(v(t)) \mid \mathscr{H}_t] \\ &= \mathbb{P}\left[\{r(t) \leq p(v(t))\} \cap \left\{ \cup_{k=1}^{N(\varepsilon)+1} E_k \right\} \mid \mathscr{H}_t \right] \\ &= \sum_{k=1}^{N(\varepsilon)+1} \mathbb{P}[\{r(t) \leq p(v(t))\} \cap E_k \mid \mathscr{H}_t].\end{aligned} \tag{3.43}$$

Next, we obtain upper and lower bounds of terms $\mathbb{P}[\{r(t) \leq p(v(t))\} \cap E_k \mid \mathscr{H}_t]$ on the far right-hand side of (3.43).

First, it follows from (3.41) that

$$\mathbb{P}[\{r(t) \leq p(v(t))\} \cap E_k \mid \mathscr{H}_t] \leq \mathbb{P}\left[\left\{ r(t) \leq \frac{k}{N(\varepsilon)} \right\} \cap E_k \mid \mathscr{H}_t \right].$$

Therefore, since $E_k \in \mathscr{H}_t$, and moreover, $r(t)$ is independent of $\mathscr{H}_t$, we have

$$\begin{aligned}\mathbb{P}[\{r(t) \le p(v(t))\} \cap E_k | \mathscr{H}_t] &\le \mathbb{P}\left[r(t) \le \frac{k}{N(\varepsilon)}\right] \mathbb{P}[E_k \mid \mathscr{H}_t] \\ &= \mathbb{P}\left[r(t) \le \frac{k}{N(\varepsilon)}\right] \mathbb{1}[E_k]. \end{aligned} \tag{3.44}$$

Since $\mathbb{P}\left[r(t) \le \frac{k}{N(\varepsilon)}\right] = \frac{k}{N(\varepsilon)}$ for $k \in \{1, \ldots, N(\varepsilon)\}$ and $\mathbb{P}\left[r(t) \le \frac{k}{N(\varepsilon)}\right] = 1 \le \frac{k}{N(\varepsilon)}$ for $k = N(\varepsilon) + 1$, we have $\mathbb{P}[r(t) \le \frac{k}{N(\varepsilon)}] \le \frac{k}{N(\varepsilon)}$. It then follows from (3.44) that

$$\mathbb{P}[\{r(t) \le p(v(t))\} \cap E_k \mid \mathscr{H}_t] \le \frac{k}{N(\varepsilon)} \mathbb{1}[E_k]. \tag{3.45}$$

Furthermore, (3.41) implies

$$\mathbb{P}[\{r(t) \le p(v(t))\} \cap E_k \mid \mathscr{H}_t] \ge \mathbb{P}\left[\left\{r(t) \le \frac{k-1}{N(\varepsilon)}\right\} \cap E_k \mid \mathscr{H}_t\right].$$

Here, by noting that $\mathbb{P}\left[r(t) \le \frac{k-1}{N(\varepsilon)}\right] = \frac{k-1}{N(\varepsilon)}$ for all $k \in \{1, \ldots, N(\varepsilon) + 1\}$, and by using techniques similar to those that we used for obtaining (3.45), we get

$$\mathbb{P}[\{r(t) \le p(v(t))\} \cap E_k \mid \mathscr{H}_t] \ge \frac{k-1}{N(\varepsilon)} \mathbb{1}[E_k]. \tag{3.46}$$

Hence, it follows from (3.43), (3.45), and (3.46) that

$$\sum_{k=1}^{N(\varepsilon)+1} \frac{k-1}{N(\varepsilon)} \mathbb{1}[E_k] \le \mathbb{E}[\mathbb{1}[r(t) \le p(v(t))] \mid \mathscr{H}_t] \le \sum_{k=1}^{N(\varepsilon)+1} \frac{k}{N(\varepsilon)} \mathbb{1}[E_k]. \tag{3.47}$$

It is shown in (3.42) and (3.47) that the random variables $p(v(t))$ and $\mathbb{E}[\mathbb{1}[r(t) \le p(v(t))] | \mathscr{H}_t]$ are lower- and upper-bounded by the same quantities. Hence, by (3.42) and (3.47)

$$\begin{aligned}|\mathbb{E}[\mathbb{1}[r(t) \le p(v(t))] \mid \mathscr{H}_t] - p(v(t))| &\le \sum_{k=1}^{N(\varepsilon)+1} \frac{k}{N(\varepsilon)} \mathbb{1}[E_k] - \sum_{k=1}^{N(\varepsilon)+1} \frac{k-1}{N(\varepsilon)} \mathbb{1}[E_k] \\ &= \frac{1}{N(\varepsilon)} \sum_{k=1}^{N(\varepsilon)+1} \mathbb{1}[E_k] = \frac{1}{N(\varepsilon)} \le \varepsilon, \end{aligned} \tag{3.48}$$

for all $\varepsilon \in (0, 1]$. Since ε is arbitrary, (3.48) implies that $\mathbb{E}[\mathbb{1}[r(t) \le p(v(t))] \mid \mathscr{H}_t] = p(v(t))$. Therefore, by (3.40), $\mathbb{E}[l(t) - p(v(t)) \mid \mathscr{F}_{t-1}] = 0$. Finally, since $M(t) = l(t) - p(v(t)) + M(t-1)$ and $M(t-1)$ is $\mathscr{F}_{t-1}$-measurable, we have $\mathbb{E}[M(t) \mid \mathscr{F}_{t-1}] = \mathbb{E}[l(t) - p(v(t)) + M(t-1) \mid \mathscr{F}_{t-1}] = M(t-1)$, which completes the proof. □

Now by using this lemma and Azuma's inequality for martingales (see [14, 35]), we obtain the following result.

Lemma 3.3 *Consider the process* $\{M(t) \in \mathbb{R}\}_{t\in\mathbb{N}_0}$ *defined in (3.38). For any* $\tilde{\varepsilon} > 0$, *we have*

$$\mathbb{P}[M(t-1) > \tilde{\varepsilon}t] \leq e^{-\tilde{\varepsilon}^2 t/2}, \quad t \in \mathbb{N}. \tag{3.49}$$

Proof By Lemma 3.2, $M(\cdot)$ is a martingale with respect to the filtration $\{\mathscr{F}_t\}_{t\in\mathbb{N}_0}$. Let the process $\{\overline{M}(t) \in \mathbb{R}\}_{t\in\mathbb{N}_0}$ be given by $\overline{M}(0) \triangleq 0$ and $\overline{M}(t) \triangleq M(t-1)$, $t \in \mathbb{N}$. With this definition, $\overline{M}(\cdot)$ is a martingale with respect to the filtration $\overline{\mathscr{F}}_t$, $t \in \mathbb{N}_0$, given by $\overline{\mathscr{F}}_0 \triangleq \{\emptyset, \Omega\}$ and $\overline{\mathscr{F}}_t \triangleq \mathscr{F}_{t-1}$, $t \in \mathbb{N}$. Note that $|\overline{M}(t) - \overline{M}(t-1)| = |M(t-1) - M(t-2)| = |l(t-1) - p(t-1)| \leq 1$ for $t \in \mathbb{N}$. By Azuma's inequality (see Sect. 3.3 of [35] and Theorem 14.12 of [14]) for martingales with bounded increments, we have

$$\mathbb{P}[\overline{M}(t) \geq \tilde{\varepsilon}t] = \mathbb{P}[\overline{M}(t) - \overline{M}(0) \geq \tilde{\varepsilon}t] \leq e^{-\tilde{\varepsilon}^2 t/2}, \quad t \in \mathbb{N},$$

which implies (3.49), since $\mathbb{P}[M(t-1) > \tilde{\varepsilon}t] \leq \mathbb{P}[M(t-1) \geq \tilde{\varepsilon}t] = \mathbb{P}[\overline{M}(t) \geq \tilde{\varepsilon}t]$. □

We remark that $M(t-1)$ in Lemma 3.3 corresponds to the difference between the sums $\sum_{i=0}^{t-1} l(i)$ and $\sum_{i=0}^{t-1} p(v(i))$. Here, $\sum_{i=0}^{t-1} l(i)$ represents the total number of transmission failures in the first t time steps and $\sum_{i=0}^{t-1} p(v(i))$ represents the sum of conditional failure probabilities for those times. Lemma 3.3 indicates that as time progresses, these quantities, when averaged over time, get close to each other with high probability. In particular, by using a similar argument to that in the proof of Lemma 3.3, we can show that an inequality similar to (3.49) holds also for $-M(t-1)$, i.e., $\mathbb{P}[-M(t-1) > \tilde{\varepsilon}t] \leq e^{-\tilde{\varepsilon}^2 t/2}$ for $t \in \mathbb{N}$. As a result, Borel–Cantelli lemma (see [5]) implies

$$\lim_{t\to\infty} \frac{1}{t}\left(\sum_{i=0}^{t-1} l(i) - \sum_{i=0}^{t-1} p(v(i))\right) = 0,$$

almost surely. In our analysis, the inequality (3.49) obtained for $M(t-1)$ is used.

We can now prove Theorem 3.1 by using Lemmas 3.1 and 3.3.

Proof of Theorem 3.1: First, note that (3.32) implies $\hat{p}(\overline{v}) < 1$. To prove this result, we show that for every $\rho \in (\hat{p}(\overline{v}), 1)$

$$\sum_{t=1}^{\infty} \mathbb{P}\left[\sum_{i=0}^{t-1} l(i) > \rho t\right] < \infty. \tag{3.50}$$

Let $\varepsilon \triangleq \rho - \hat{p}(\overline{v})$. Observe that since $p(v) \leq \hat{p}(v)$, $v \in [0, \infty)$

$$\begin{aligned}\mathbb{P}\left[\sum_{i=0}^{t-1} l(i) > \rho t\right] &= \mathbb{P}\left[\sum_{i=0}^{t-1}(l(i) - p(v(i))) + \sum_{i=0}^{t-1} p(v(i)) - \hat{p}(\overline{v})t > \varepsilon t\right] \\ &\leq \mathbb{P}\left[\sum_{i=0}^{t-1}(l(i) - p(v(i))) + \sum_{i=0}^{t-1} \hat{p}(v(i)) - \hat{p}(\overline{v})t > \varepsilon t\right] \\ &\leq \mathbb{P}\left[M(t-1) > \frac{\varepsilon}{2}t\right] + \mathbb{P}\left[\sum_{i=0}^{t-1} \hat{p}(v(i)) - \hat{p}(\overline{v})t > \frac{\varepsilon}{2}t\right],\end{aligned}$$

and hence

$$\sum_{t=1}^{\infty}\mathbb{P}\left[\sum_{i=0}^{t-1} l(i) > \rho t\right] \leq \sum_{t=1}^{\infty}\mathbb{P}\left[M(t-1) > \frac{\varepsilon}{2}t\right] + \sum_{t=1}^{\infty}\mathbb{P}\left[\sum_{i=0}^{t-1} \hat{p}(v(i)) - \hat{p}(\overline{v})t > \frac{\varepsilon}{2}t\right], \tag{3.51}$$

where $M(\cdot)$ is given by (3.37). By Lemma 3.3 with $\tilde{\varepsilon} = \varepsilon/2$

$$\sum_{t=1}^{\infty}\mathbb{P}\left[M(t-1) > \frac{\varepsilon}{2}t\right] \leq \sum_{t=1}^{\infty} e^{-\varepsilon^2 t/8} = \frac{e^{-\varepsilon^2/8}}{1 - e^{-\varepsilon^2/8}} < \infty. \tag{3.52}$$

Now, we show $\sum_{t=1}^{\infty}\mathbb{P}[\sum_{i=0}^{t-1}\hat{p}(v(i)) - \hat{p}(\overline{v})t > \frac{\varepsilon}{2}t] < \infty$. It follows from Lemma 3.1 that for $\tilde{\varepsilon} = \varepsilon/2$, there exists $\tilde{\tau} \in \mathbb{N}$ such that (3.34) holds. If $\tilde{\tau} = 1$, then by using (3.34), we obtain $\sum_{t=1}^{\infty}\mathbb{P}[\sum_{i=0}^{t-1}\hat{p}(v(i)) - \hat{p}(\overline{v})t > \frac{\varepsilon}{2}t] = 0$. If, on the other hand, $\tilde{\tau} > 1$, then (3.34) implies

$$\begin{aligned}\sum_{t=1}^{\infty}\mathbb{P}\left[\sum_{i=0}^{t-1} \hat{p}(v(i)) - \hat{p}(\overline{v})t > \frac{\varepsilon}{2}t\right] &= \sum_{t=1}^{\tilde{\tau}-1}\mathbb{P}\left[\sum_{i=0}^{t-1} \hat{p}(v(i)) - \hat{p}(\overline{v})t > \frac{\varepsilon}{2}t\right] \\ &\leq \tilde{\tau} - 1 < \infty.\end{aligned} \tag{3.53}$$

Finally, it follows from (3.51)–(3.53) that (3.50) holds for all $\rho \in (\hat{p}(\overline{v}), 1)$. Now by (3.50) and Lemma 3.3 of [6], we have

$$\limsup_{t\to\infty}\frac{1}{t}\sum_{i=0}^{t-1} l(i) \leq \hat{p}(\overline{v}), \tag{3.54}$$

almost surely. Hence, the result follows from Theorem 3.5 of [6] with $\rho = \hat{p}(\overline{v})$. □

We remark that Theorem 3.1 allows us to use the bound $\overline{v}$ of the average interference powers to analyze the effect of jamming attacks. The quantity $\overline{v}$ is directly related to the average energy of the attacker. Notice that κ from Assumption 3.1 is not present in the stability conditions, since it only affects the initial behavior of the attacker. Although asymptotic stability properties do not depend on κ, the state trajectories for finite durations are affected by how large κ is. Attackers satisfying Assumption 3.1 with large κ values can cause more failures during the initial time steps.

In the next section, the efficacy of Theorem 3.1 is illustrated for an example system under the rolling-horizon jamming attacks discussed in Sect. 3.3.

3.5 Numerical Example

In this section, we illustrate our results for the state-dependent jamming attacks against a networked control system (3.1), (3.2) with

$$A = \begin{bmatrix} 0.1 & -1 \\ 1.1 & 1.8 \end{bmatrix}, \quad B = \begin{bmatrix} 0 \\ 1 \end{bmatrix}, \quad K = [-0.9277 \; - 1.2615],$$

and the wireless channel characterized with $p(\cdot)$ given by (3.4) where $c = 1, \xi = 3$, and $\sigma = 0.4$. Previously in [7], we considered the effects of time-dependent attacks on this system. Here, our goal is to demonstrate the effects of state-dependent rolling-horizon attacks discussed in Sect. 3.3 as well as the utility of our closed-loop system stability results presented in Sect. 3.4.

To illustrate the utility function in the rolling-horizon attacks, we first consider the simple case with a short horizon length $T_{\mathrm{A}} = 2$ together with the parameters $\kappa = 0$ and $\overline{v} = 3.5$. For this case, the initial utility function $U(x_0, v_0, v_1)$ is shown in Fig. 3.4 for two different values of x_0. The utility shape is quite different for the two cases. It is interesting to note that in the top plot, the attack level v_0 at the first time step $t = 0$ is more critical and should be larger while the attack at time $t = 1$ has relatively small effect. In contrast, in the bottom plot, the attacker obtains a larger utility if the interference power v_1 for the second time step $t = 1$ is selected large. As there is an average power constraint (Assumption 3.1), choosing large values for both v_0 and v_1 is not possible, and the difference in the shape of utilities (due to different state values) plays a critical role in the interference power choices of the attacker at each time.

By Proposition 3.1, the utility $U(x_0, v_0, v_1)$ is directly related to the expected second moments of future states $x(1)$ and $x(2)$. The attacker uses the current state information $x(0) = x_0$ and attempts to maximize this utility while keeping the interference powers within certain constraint sets. The left part of Fig. 3.5 provides the contour plot of the utility function $U(x_0, v_0, v_1)$ with $x_0 \triangleq [1.0, \; 0.85]^{\mathrm{T}}$, where each labeled curve indicates a set of interference power pairs (v_0, v_1) that result in the same utility value indicated in the label. The interference power constraints are represented with the orange shaded region, and the optimal value (v_0^*, v_1^*) that is obtained by numerically maximizing the utility is shown with a red point. The attacker sets the first interference power as $v(0) = v_0^*$. Notice that the state at the next time instant depends on whether the transmission is successful ($l(0) = 0$) or it fails ($l(0) = 1$). As a result, the utility function $U(x_1, v_1, v_2)$ for time $t = 1$ may take two different

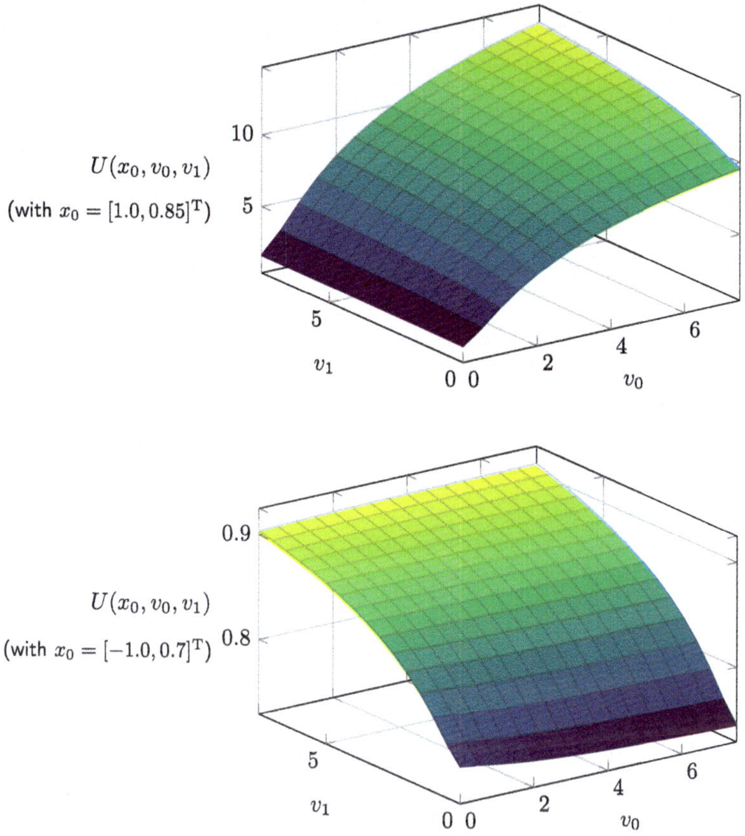

Fig. 3.4 Attacker's utility $U(x_0, v_0, v_1)$ with respect to possible interference powers v_0, v_1 for given state values $x_0 \triangleq [1.0,\ 0.85]^{\mathrm{T}}$ (top) and $x_0 \triangleq [-1.0,\ 0.7]^{\mathrm{T}}$ (bottom)

forms as indicated in the right part of Fig. 3.5. Notice that the optimal interference power levels picked by the attacker in these two cases are different.

For the average interference power bound $\overline{v} = 3.5$ considered above, Theorem 3.1 implies that the closed-loop system is almost surely asymptotically stable. To show stability, we note that $\beta = 0.3$, $\varphi = 5.4884$, and

$$P = \begin{bmatrix} 0.7728 & 0.8554 \\ 0.8554 & 3.2649 \end{bmatrix} \tag{3.55}$$

satisfy the conditions (3.30)–(3.32) of Theorem 3.1 with the continuous, nondecreasing, and concave function $\hat{p}(v) \triangleq p(v + (c\xi - 3\sigma)/3)$.

In Fig. 3.6, we show 100 sample state-norm trajectories obtained for different κ and $\overline{v}$ values. When κ is large, the state magnitude can reach larger values, since large κ implies large initial attack resources that the attacker can utilize. The parameter $\overline{v}$

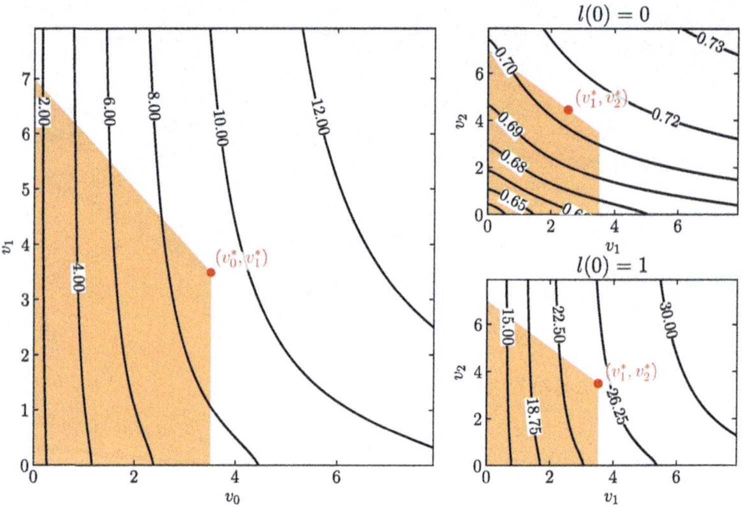

Fig. 3.5 Contour plots of the attacker's utility functions for times $t = 0$ (left) and $t = 1$ (right) with respect to the possible actions of attacker

represents a bound on the attacker's average interference power and it has a bigger effect on system stability. When $\overline{v} = 3.5$, as guaranteed by Theorem 3.1, the state trajectories converge to the origin. On the other hand when $\overline{v} = 10$, stability is no longer guaranteed by Theorem 3.1. The simulations for this case indicate that the state magnitude may grow to much larger values. Figure 3.7 shows a sample realization of the attacker's interference power following the rolling-horizon strategy with $\kappa = 0$ and $\overline{v} = 3.5$. The interference power deviates from the average interference power at certain times; however, Assumption 3.1 holds.

In Fig. 3.8, we compare three attack strategies: (i) constant-power attack strategy where $v(t) = 3.5$, $t \in \mathbb{N}_0$, (ii) state-dependent rolling-horizon strategy with $\kappa = 0$, $\overline{v} = 3.5$, $T_{\mathrm{A}} = 2$, and (iii) state-dependent rolling-horizon strategy with $\kappa = 0$, $\overline{v} = 3.5$, $T_{\mathrm{A}} = 4$. In all cases, the average interference power is 3.5, and stability is guaranteed. The effects of these different attack strategies are compared by the values of the second moment sum $\sum_{i=0}^{t} \mathbb{E}[\|x(i)\|^2]$ approximated over 100 samples. We note that $\sum_{i=0}^{t} \mathbb{E}[\|x(i)\|^2]$ takes larger values when the attacks follow the state-dependent rolling-horizon strategy. Within the state-dependent cases, the value is even larger for the case with larger optimization horizon T_{A}. This indicates that an attacker with larger computational resources can use jamming resources more efficiently and be more damaging.

To check the robustness of the above-mentioned three strategies against noise in state measurements, we also consider an additional scenario. In this scenario, both the attacker and the controller have access to noisy state measurements $x(t) + w(t)$ at time t, where $w(0), w(1), \ldots$ are independent random variables that are distributed uniformly in $[-0.2, 0.2]$. The second moment sum $\sum_{i=0}^{t} \mathbb{E}[\|x(i)\|^2]$ obtained for

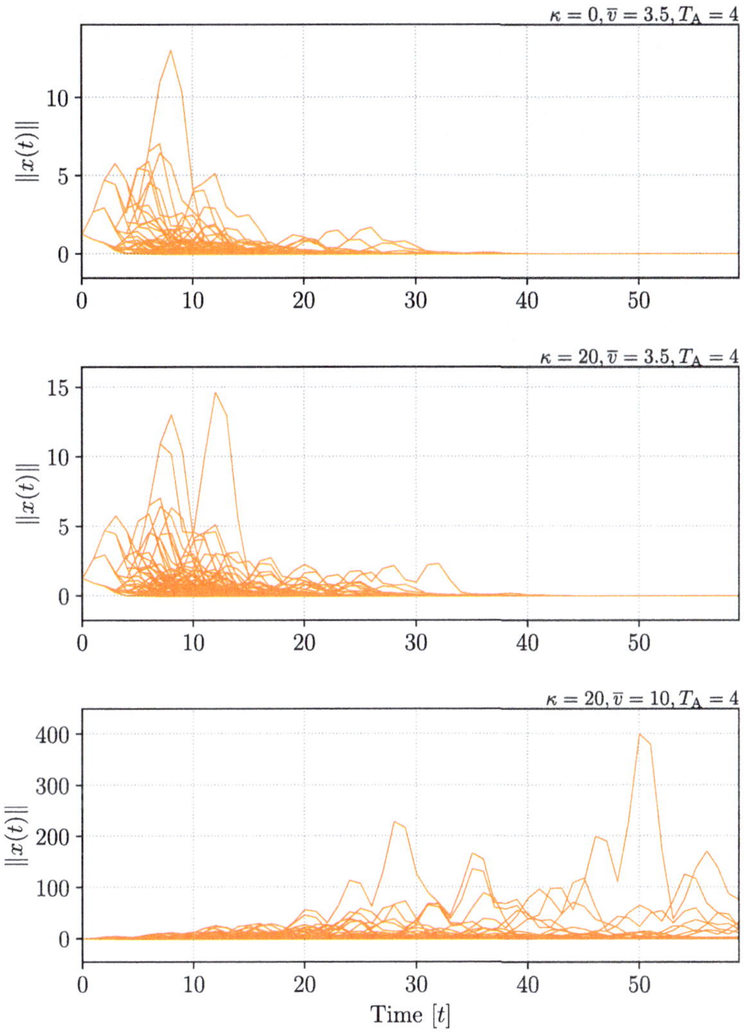

Fig. 3.6 Comparison of the sample paths of state norm for different values of parameters κ and $\overline{v}$

different attack strategies are shown in Fig. 3.9. Notice that in this case, convergence is not achieved due to noise and $\sum_{i=0}^{t} \mathbb{E}[\|x(i)\|^2]$ grows over time unboundedly. Figure 3.9 indicates that the rolling-horizon strategy is still more damaging compared to the constant-power strategy. It is, however, interesting to note that if the controller has access to noise-free measurements, but the attacker's observation of the state is noisy, then state-dependent rolling-horizon strategy does not perform well when the noise variance is large. This is because the optimal interference powers obtained for the noisy state measurements may not be optimal for the actual states.

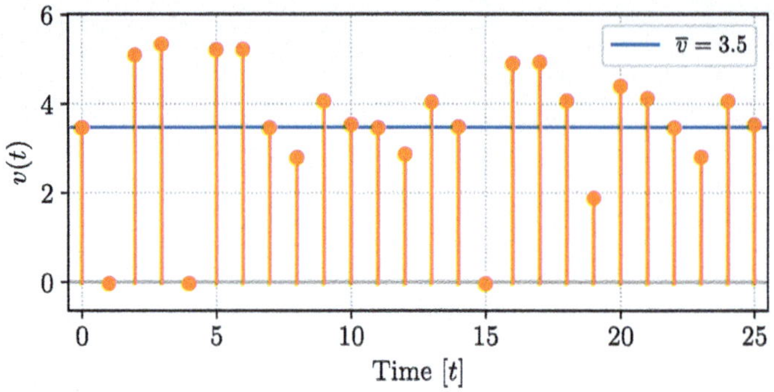

Fig. 3.7 Interference powers with respect to time

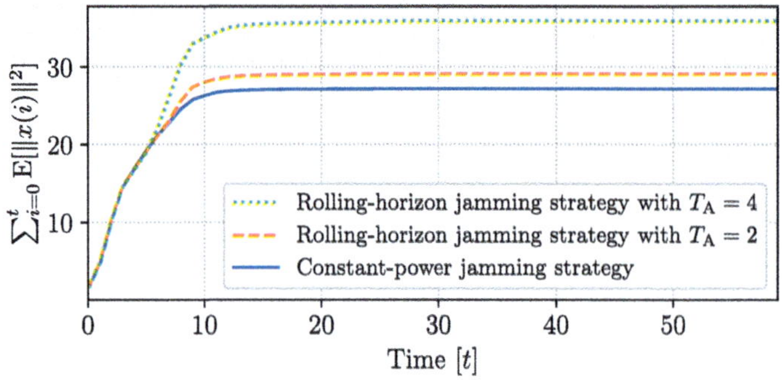

Fig. 3.8 Comparison of the effects of different attack strategies

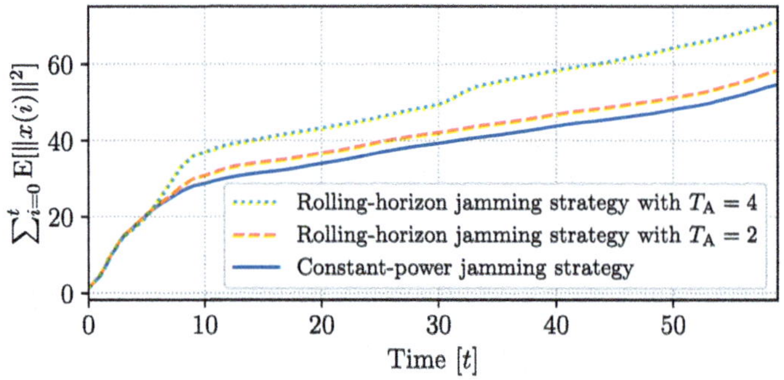

Fig. 3.9 Comparison of the effects of different attack strategies when the state measurements are noisy

3.6 Conclusion

In this chapter, we explored the problem of controlling a linear plant over a vulnerable wireless channel. After giving a general introduction to networked control under jamming attacks, we investigated the scenarios where the jamming attacker can use the state information to decide the jamming interference power at each time. In particular, we looked at rolling-horizon attacks, where the attacker uses the present state information to maximize a certain utility function that involves predicted future states. In our wireless channel model, the likelihood of a failure in transmission of the control input packets depends on the interference power selected by the attacker. The particular interference powers used by the attacker at each time are assumed to be unknown since they are part of the attacker's strategy. For this problem setting, we obtained sufficient networked control system stability conditions in terms of the upper bound of the attacker's average interference power, which is related to the energy constraints of the attacker.

There are several potential future research directions. For instance, designing control and communication mechanisms with the intention of mitigating jamming attacks is a promising direction. Here in this chapter, we looked at the problem of optimizing jamming interference from the perspective of an attacker. It is possible to extend this result by also investigating optimization of the power used for wireless transmission of the control data. Essentially, this optimization problem is to be solved by the system operator to achieve better protection against jamming.

The security problem with an optimization setting naturally leads to a game-theoretic formulation, where both the system operator and the attacker attempt to optimize their strategies at the same time. There are a few works that tackle this problem using different approaches for different formulations [1, 4, 23, 30, 49]. The frameworks developed there can be extended to investigate the wireless networked control problem while also incorporating energy costs and constraints of the players.

Another promising area for future research concerns multi-agent systems under jamming attacks. Especially, SINR-based formulations in multi-agent consensus and distributed control/computation problems could be useful from a practical viewpoint considering future applications in IoT.

Acknowledgements This work is supported by JSPS KAKENHI Grant Numbers 20K14771 and 18H01460, JST ERATO HASUO Metamathematics for Systems Design Project (No. JPMJER1603), and by JST CREST Grant No. JPMJCR15K3.

References

1. Alpcan, T., Başar, T.: Network Security: A Decision and Game-Theoretic Approach. Cambridge University Press (2010)
2. Amini, M., Usunier, N.: Learning with Partially Labeled and Interdependent Data. Springer (2015)

3. Bemporad, A., Heemels, M., Johansson, M. (eds.): Networked Control Systems. Lecture Notes in Control and Information Sciences, vol. 406. Springer (2010)
4. Bhattacharya, S., Gupta, A., Başar, T.: Jamming in mobile networks: a game-theoretic approach. J. Num. Algeb. Control Optim. **3**(1), 1–30 (2013)
5. Billingsley, P.: Probability and Measure. Wiley (2012)
6. Cetinkaya, A., Ishii, H., Hayakawa, T.: Networked control under random and malicious packet losses. IEEE Trans. Autom. Control **62**(5), 2434–2449 (2017)
7. Cetinkaya, A., Ishii, H., Hayakawa, T.: The effect of time-varying jamming interference on networked stabilization. SIAM J. Control Optim. **56**, 2398–2435 (2018)
8. Cetinkaya, A., Ishii, H., Hayakawa, T.: A probabilistic characterization of random and malicious failures communication failures in multi-hop networked control. SIAM J. Control Optim. **56**, 2398–2435 (2018)
9. Cetinkaya, A., Ishii, H., Hayakawa, T.: State-dependent jamming interference in networked stabilization. In: Proceedings of IEEE Conference on Decision and Control, pp. 7249–7254 (2018)
10. Cetinkaya, A., Ishii, H., Hayakawa, T.: An overview on denial-of-service attacks in control systems: attack models and security analyses. Entropy **21**(2) (2019). Article no. 210
11. Cetinkaya, A., Kikuchi, K., Hayakawa, T., Ishii, H.: Randomized transmission protocols for protection against jamming attacks in multi-agent consensus. Automatica **117** (2020)
12. Cheng, P., Shi, L., Sinopoli, B. (eds.): Special Issue on Secure Control of Cyber Physical Systems, IEEE Trans. Control Netw. Syst. **4** (2017)
13. Chong, M.S., Sandberg, H., Teixeira, A.M.H.: A tutorial introduction to security and privacy for cyber-physical systems. In: Proceedings of European Control Conference, pp. 968–978 (2019)
14. DasGupta, A.: Probability for Statistics and Machine Learning. Springer (2011)
15. De Persis, C., Tesi, P.: Input-to-state stabilizing control under Denial-of-Service. IEEE Trans. Autom. Control **60**(11), 2930–2944 (2015)
16. De Persis, C., Tesi, P.: Networked control of nonlinear systems under Denial-of-Service. Syst. Control Lett. **96**, 124–131 (2016)
17. Dibaji, S.M., Pirani, M., Flamholz, D.B., Annaswamy, A.M., Johansson, K.H., Chakrabortty, A.: A systems and control perspective of CPS security. Ann. Rev. Contr. pp. 394–411 (2019)
18. Durrett, R.: Probability: Theory and Examples. Cambridge University Press (2010)
19. Fawzi, H., Tabuada, P., Diggavi, S.: Secure estimation and control for cyber-physical systems under adversarial attacks. IEEE Trans. Autom. Control **59**(6), 1454–1467 (2014)
20. Feng, S., Tesi, P.: Resilient control under Denial-of-Service: Robust design. Automatica **79**, 42–51 (2017)
21. Fragkiadakis, A., Askoxylakis, I., Chatziadam, P.: Denial-of-service attacks in wireless networks using off-the-shelf hardware. In: Proceedings of International Conference on Distributed, Ambient and Pervasive Interactions, pp. 427–438 (2014)
22. Goldsmith, A.: Wireless Communications. Cambridge University Press (2005)
23. Gupta, A., Nayyar, A., Langbort, C., Başar, T.: A dynamic transmitter-jammer game with asymmetric information. In: Proceedings of IEEE Conference on Decision and Control, pp. 6477–6482 (2012)
24. Gupta, V., Martins, N.C., Baras, J.S.: Optimal output feedback control using two remote sensors over erasure channels. IEEE Trans. Autom. Control **54**(7), 1463–1476 (2009)
25. Hespanha, J.P., Naghshtabrizi, P., Xu, Y.: A survey of recent results in networked control systems. Proc. IEEE **95**(1), 138–172 (2007)
26. Ishii, H.: H-infinity control with limited communication and message losses. Syst. Control Lett. **57**, 322–331 (2008)
27. Jhaveri, R.H., Patel, S.J., Jinwala, D.C.: DoS attacks in mobile ad hoc networks: a survey. In: Proceedings of ACTT, pp. 535–541 (2012)
28. Just, M., Kranakis, E., Wan, T.: Resisting malicious packet dropping in wireless ad hoc networks. In: Proceedings of International Conference on Ad-Hoc Network Wireless, pp. 151–163 (2003)

29. Lemmon, M., Hu, X.S.: Almost sure stability of networked control systems under exponentially bounded bursts of dropouts. In: Proceedings of 14th HSCC, pp. 301–310 (2011)
30. Li, Y., Quevedo, D.E., Dey, S., Shi, L.: SINR-based DoS attack on remote state estimation: a game-theoretic approach. IEEE Trans. Control Netw. Syst. **4**(3), 632–643 (2017)
31. Li, Y., Shi, L., Cheng, P., Chen, J., Quevedo, D.E.: Jamming attacks on remote state estimation in cyber-physical systems: a game-theoretic approach. IEEE Trans. Autom. Control **60**(10), 2831–2836 (2015)
32. Lipp, T., Boyd, S.: Antagonistic control. Syst. Control Lett. **98**, 44–48 (2016)
33. Lun, Y.Z., D'Innocenzo, A., Smarra, F., Malavolta, I., Di Benedetto, M.D.: State of the art of cyber-physical systems security: an automatic control perspective. J. Syst. Softw. **149**, 174–216 (2019)
34. Mahmoud, M.M.E.A., Shen, X.S.: Security for Multi-hop Wireless Networks. Springer (2014)
35. McDiarmid, C.: Concentration. In: Probabilistic Methods for Algorithmic Discrete Mathematics, pp. 195–248. Springer (1998)
36. Mo, Y., Garone, W., Casavola, A., Sinopoli, B.: False data injection attacks against state estimation in wireless sensor networks. In: Proceedings of IEEE Conference on Decision Control, pp. 5967–5972 (2010)
37. Nair, C.: Upper concave envelopes and auxiliary random variables. Int J. Adv. Eng. Sci. Appl. Math. **5**(1) (2013)
38. Øksendal, B.: Stochastic Differential Equations. Springer (2005)
39. Pelechrinis, K., Iliofotou, M., Krishnamurty, S.V.: Denial of service attacks in wireless networks: the case of jammers. IEEE Commun. Surveys Tuts. **13**(2), 245–257 (2011)
40. Quevedo, D.E., Østergaard, J., Nešić, D.: Packetized predictive control of stochastic systems over bit-rate limited channels with packet loss. IEEE Trans. Autom. Control **56**(12), 2854–2868 (2011)
41. Rudin, W.: Principles of Mathematical Analysis. McGraw-Hill (1976)
42. Sandberg, H., Amin, S., Johansson, K.H.: Special issue on cyberphysical security in networked control systems. IEEE Control Syst. Mag. **35**(1) (2015)
43. Senejohnny, D., Tesi, P., De Persis, C.: A jamming-resilient algorithm for self-triggered network coordination. IEEE Trans. Control Netw. Syst. **5**(3), 981–990 (2017)
44. Shi, G., Li, K.: Signal Interference in WiFi and ZigBee Networks. Springer (2017)
45. Shisheh Foroush, H., Martínez, S.: On triggering control of single-input linear systems under pulse-width modulated DoS signals. SIAM J. Control Optim. **54**(6), 3084–3105 (2016)
46. Xie, W.C.: Dynamic Stability of Structures. Cambridge University Press (2011)
47. Xu, W., Trappe, W., Zhang, Y., Wood, T.: The feasibility of launching and detecting jamming attacks in wireless networks. In: Proceedings of ACM International Symposium on Mobile Ad Hoc Networking and Computing, pp. 46–57 (2005)
48. Zhang, H., Qi, Y., Wu, J., Fu, L., He, L.: DoS attack energy management against remote state estimation. IEEE Trans. Control Netw. Syst. **5**, 383–394 (2018)
49. Zhu, Q., Saad, W., Han, Z., Poor, H.V., Başar, T.: Eavesdropping and jamming in next-generation wireless networks: a game-theoretic approach. In: Proceedings of IEEE Military Communications Conference, pp. 119–124 (2011)

Part II
Secure Communication and Privacy in Control

Chapter 4
Resilient Hierarchical Networked Control Systems: Secure Controls for Critical Locations and at Edge

Azwirman Gusrialdi and Zhihua Qu

Abstract Integration of information and communication technology (ICT) offers new opportunities in improving the management and operation of critical infrastructures such as power systems as it allows connection of different sensors and control components via a communication network, leading to the so-called networked control systems (NCS). However, the use of open and pervasive ICT such as the Internet or wireless communication technologies comes at a price of making NCS vulnerable to cyber intrusions/attacks which may cause physical damage. This chapter presents control algorithms to ensure resilient and safe operation of NCS under unknown cyberattacks. Specifically, a variant of dynamic watermarking strategies is presented by embedding encoding/decoding components of chaotic signals into the NCS for secure control for critical locations where the measurement/control signals are transmitted to/from the control center via a communication network. In addition, resilient cooperative control algorithms are discussed to ensure safe operation at edge of the NCS which consists of a large number of distributed controllable devices. Several numerical examples are provided to illustrate the proposed control strategies.

4.1 Introduction

Information and communication technology (ICT) has been increasingly integrated and deployed into critical infrastructures (i.e., physical systems) such as power systems and transportation systems. ICT offers new opportunities in improving the management and operation of the critical infrastructures. In particular, ICT (also called

A. Gusrialdi (✉)
Faculty of Engineering and Natural Sciences, Tampere University,
33014 Tampere, Pirkanmaa, Finland
e-mail: azwirman.gusrialdi@tuni.fi

Z. Qu
Department of Electrical and Computer Engineering, University of Central Florida,
Orlando, FL 32816, USA
e-mail: qu@ucf.edu

H. Ishii and Q. Zhu (eds.), *Security and Resilience of Control Systems*,
Lecture Notes in Control and Information Sciences 489,
https://doi.org/10.1007/978-3-030-83236-0_4

cyber-layer) provides an efficient way and great flexibility in sharing information between sensors for monitoring purposes and controlling geographically distributed small devices over a communication network which is also known as networked control systems (NCS).

In practice, the control of large-scale complex systems such as power grid is performed in a hierarchical manner. Critical locations are controlled remotely at the control station/center, that is the measurement/control signals are transmitted to/from the control station via a communication network. For example, system protection, i.e., protective relaying is a very crucial component for ensuring the reliability of power system operation. Advanced protective relay schemes involve telecommunications for monitoring and remotely tripping or controlling the breakers of critical equipment in order to clear power faults [2]. Another example is wide-area monitoring and control of power system using real-time measurements from phasor measurement units (PMUs) [4]. To this end, each generator sends its measurement to a cloud or virtual machine which computes the control input. The control inputs are then sent back to the generator for damping the wide-area oscillations. On the other hand, controls at the edge of critical infrastructures call for scalable decision-making algorithms to deal with the large number of small and distributed controllable devices. For example, as more Distributed Energy Resources (DERs)—distributed generation, batteries, and controllable loads—are being integrated into the distribution network of future power system, centralized control algorithm where all the data processing and computation are performed at a control center/station will need to be replaced by distributed control algorithm performed by each individual DER by exchanging local information via a communication network to compute their decisions. Distributed optimization and control algorithms have several potential advantages over centralized approaches including scalability to the system's size, robustness with respect to failure of individual agent, and preserving data privacy [27]. It is evident that ICT plays an important role in realizing efficient controls for critical locations and at the edge of the critical infrastructures.

The use of open and pervasive ICT such as the Internet or wireless communication technologies comes at a price of making NCS vulnerable to cyber intrusions/attacks which may cause physical damage to the critical infrastructure due to the tight coupling between the physical system and cyber layer [8, 22]. An example of cyber-attacks on critical infrastructure is the synchronized and coordinated attack on the Ukraine power grid in 2015 [29], causing a 6 hour blackout and affecting hundreds of thousands of customers. Unfortunately, traditional security solutions in the ICT domain, focusing on data secrecy, integrity, and availability, are not sufficient to ensure the security of NCS [34]. System and control theories have shown promises in analyzing and ensuring security of NCS against intelligent attacks by leveraging fundamental understanding of the physical system dynamics and its interconnection with the cyber components [5]. In this chapter, secure controls for critical locations and at edge of NCS will be presented and discussed from the perspective of system and control theory.

Traditionally, secure control for critical locations is realized by implementing detection algorithm to investigate whether the system is healthy or operating abnormally. A standard and classical detection method is the residual-based detector scheme which is also known as passive detection [6]. However, such detection scheme is undermined by coordinated and stealthy network attacks launched by sophisticated/intelligent adversaries [8]. Recently, there have been efforts in developing active detection methods to detect stealthy attacks launched by intelligent adversaries. The idea is to exploit the knowledge of physical system and alter the system's input or dynamics to detect stealthy attacks [34]. An example of active detection method is physical watermarking achieved by inserting a noisy control input on top of the optimal input of a system [26, 31]. Since the measurement is correlated to the physical watermark through the system dynamics, the absence of watermark in the system output is an indication of a faulty behavior. Other active detection methods include a moving target approach by modifying system matrices or adding additional dynamics [32, 33]. However, these methods may increase installation costs and degrade control performance of NCS in the absence of attacks. Section 4.2 of this chapter presents a variant of dynamic watermarking-based methods to ensure secure control for critical locations. To this end, encoding/decoding components of chaotic signals are embedded into the NCS and designed to detect any stealthy system integrity attacks and also preserve control performance of the NCS in the absence of attacks.

Secure distributed control at the edge of critical infrastructure introduces new challenges due to the system's size and the interaction between the subsystems via the physical interconnection and/or the communication network which prevents the implementation of the previously mentioned approaches. Most of the results on secure distributed control are based on detecting and identifying the attacks followed by isolating the compromised subsystems, see for example [17, 21, 28, 39]. However, the strategies have limitations on communication network topology, type of attacks (attacker's strategy) or number of compromised nodes and distributed control problem under consideration. Furthermore, one important issue that needs to be considered in all of these approaches is that the stability of the system may already have been compromised before the attack is detected. Since cyberattacks cannot be foreseen in advance, it is, therefore, desirable to design distributed control algorithms so that the overall system becomes resilient against unknown attacks. Such algorithms which are capable of maintaining or restoring systems performance under unexpected events are commonly referred to as resilient control algorithms. Examples of resilient control strategies in power grid include removing/neglecting compromised data [1] and self-organizing of communication architecture [3] in order to mitigate the adverse effects of attacks. While resilient control algorithm is able to ensure the boundedness of physical variables under attacks, operating point of the compromised system may still violate the operational constraint. Hence, it is necessary to guarantee both the resilience and safety of the NCS. In Sect. 4.3 of this chapter, distributed control algorithms to ensure the resilient and safe operation of NCS are presented. Specifically, a virtual system interconnected with the distributed control algorithms originally designed under nominal operation of NCS is introduced. The virtual system acts as an anchor which maintains the NCS to operate around its nominal operating point under unknown attacks. Details of the results presented in this chapter including their analysis and proofs can be found in [9, 10, 12, 14, 19].

4.2 Resilient Control for Critical Locations

Consider a networked control system where the physical plant is monitored and controlled remotely via a network by a control center as illustrated in Fig. 4.1.

Specifically, dynamics of the physical plant is given by the following linear time-invariant systems:

$$\dot{x} = Ax + B\overline{u}, \quad y = Cx, \tag{4.1}$$

where $x, \overline{u}, y$ denote respectively the plant state, actual control signal applied by the plant, and the plant output/measurement. The measurement is sent to a control center via a communication network. The control center then computes the input signal u which will be sent back to the plant via a network. Furthermore, it is assumed that the triplet (A, B, C) is controllable and observable. The control center implements the following observer-based state feedback control:

$$\dot{\hat{x}} = A\hat{x} + Bu + L(\overline{y} - \hat{y}), \quad \hat{y} = C\hat{x}, \quad u = -K\hat{x}, \tag{4.2}$$

where $\hat{x}$ denotes the state estimate, $\overline{y}, \hat{y}$ are the measurement received and the output estimate respectively. Moreover gain matrices K, L are chosen such that matrices $(A - BK)$ and $(A - LC)$ are Hurwitz and the desired performance is achieved in the absence of attacks.

In practice, the communication network may be subject to (unknown) attacks. In other words, the control signal and measurement transmitted via the network may

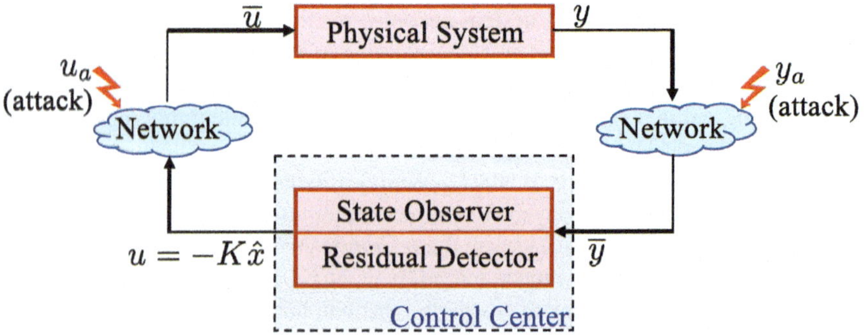

Fig. 4.1 Network control system and potential cyberattack

be corrupted given that the attacker gains access to the communication network. To reflect this situation, we write both the received measurement and control signals as

$$\overline{y} = y + y_a, \quad \overline{u} = u + u_a, \tag{4.3}$$

where y_a, u_a are the output and input vectors injected to the communication network by the attacker, respectively. In order to detect the presence of such intrusions, the control center is normally equipped with a detection algorithm which detects abnormality such as bad data. For example, the following fault detector based on threshold test [8] over the output residue $z = \overline{y} - \hat{y}$ is widely employed.

$$\begin{cases} \|Wz\| \leq \gamma & \text{normal operation}, \\ \|Wz\| > \gamma & \text{abnormal operation}, \end{cases} \tag{4.4}$$

where W is a weighting matrix and $\gamma \geq 0$ is a threshold chosen by the control center operator.

The adversary aims at launching *stealthy* system integrity attack, that is to steer and manipulate the state of the physical plant to any state that she/he wants (i.e., unsafe state) without being detected by residual detector (4.4), i.e., $\|Wz\| \leq \gamma$. In addition, if $z - z_n = 0$ where z_n denotes the output residual of the nominal system, i.e., $z_n = (y_n - \hat{y}_n)$ where $y_n/\hat{y}_n$ is the nominal system/observer output of the un-attacked control system, then the attack is called *perfectly stealthy*. On the other hand, the control center aims at detecting the presence of such perfectly stealthy attack and further ensuring resiliency of the networked control system against such attack. In the following, we will demonstrate how the adversary launches perfectly stealthy system integrity attack and how the control center could detect the presence of such attack.

4.2.1 Launching Stealthy Attack

In this subsection, we will discuss how the adversary can launch perfectly stealthy system integrity attack. First, let us assume that the adversary has the information of matrices A, B, and C in model (4.1) and he/she has gained access to the communication network to inject signals u_a, y_a in (4.3). Next, consider the case where the attack vectors u_a, y_a are generated by

$$\begin{aligned} \dot{x}_a &= Ax_a + Bu_a, \quad x_a(t_0) = 0, \\ u_a &= -K_a x_a + Hy + r_a, \quad y_a = -Cx_a \end{aligned} \tag{4.5}$$

with $r_a(t) = 0$ for $t \in [0, t_0]$ and for some starting time t_0. Then, the overall dynamics of (4.1) and (4.2) under attack (4.5) and together with the nominal overall dynamics become

$$\begin{cases} \dot{x} = Ax + B(-K\hat{x} - K_a x_a + Hy + r_a) \\ \dot{x}_a = Ax_a + B(-K_a x_a + Hy + r_a) \\ \dot{\hat{x}} = A\hat{x} - BK\hat{x} + L(y + y_a - \hat{y}) \\ \dot{x}_n = Ax_n - BK\hat{x}_n \\ \dot{\hat{x}}_n = A\hat{x}_n - BK\hat{x}_n + L(y_n - \hat{y}_n) \end{cases} \quad \begin{cases} y = Cx \\ y_a = -Cx_a \\ \hat{y} = C\hat{x} \\ y_n = Cx_n \\ \hat{y}_n = C\hat{x}_n \end{cases}$$

Applying the state transformation

$$x \longrightarrow x' \triangleq x - x_a, \quad \hat{x} \longrightarrow e_a \triangleq x' - \hat{x}, \quad \hat{x}_n \longrightarrow e_n \triangleq x_n - \hat{x}_n,$$

the state space model becomes

$$\begin{cases} \dot{x}' = (A - BK)x' + BKe_a \\ \dot{x}_a = (A - BK_a + BHC)x_a + BHCx' + Br_a \\ \dot{e}_a = (A - LC)e_a \\ \dot{x}_n = (A - BK)x_n + BKe_n \\ \dot{e}_n = (A - LC)e_n \end{cases}$$

Consider now the vector $w = e_a - e_n$. It follows that its dynamics are governed by

$$\dot{w} = (A - LC)w, \quad z - z_n = Cw,$$

with $w(t_0) = 0$. It can be observed that since matrix $(A - LC)$ is Hurwitz, we have $w(t) \equiv 0$ and consequently $z(t) = z_n(t)$ holds for the observations for all $t \in [0, \infty)$ regardless whether there is an attack or not. In other words, the adversary successfully injects system attack vectors u_a, y_a into the network while evading any residual-based detection, i.e., the attack is perfectly stealthy. Intuitively speaking, the adversary exploits the linearity of the system and by using the knowledge of the physical system, he/she is capable of designing the output attack vector y_a to cancel out the perturbation on the plant output induced by the input attack vector u_a.

In order to implement perfectly stealthy attack (4.5), in addition to system matrices A, B, and C, the adversary also needs to know real-time measurement of the plant output. However, no information of the control input or any other information at the control center, e.g., K and L is required to launch the attack. The adversary can not only modify the equilibrium point of the plant state, but also manipulate the dynamic response of the physical plant by means of exogenous injection r_a and by choosing matrices K_a and H. For example, in order to make the dynamics of the overall system unstable while keeping the attacker's own model stable by itself, the attacker can choose matrices K_a, H in (4.5) such that $(A - BK_a)$ is stable but $(A - BK_a + BHC)$ is unstable. This can be done with the knowledge of the plant (i.e., matrices A, B, and C only), and is independent of the control design (matrix K).

Next, the perfectly stealthy attack described previously is illustrated using simulation on quadruple-tank process with linearized system model described in [18].

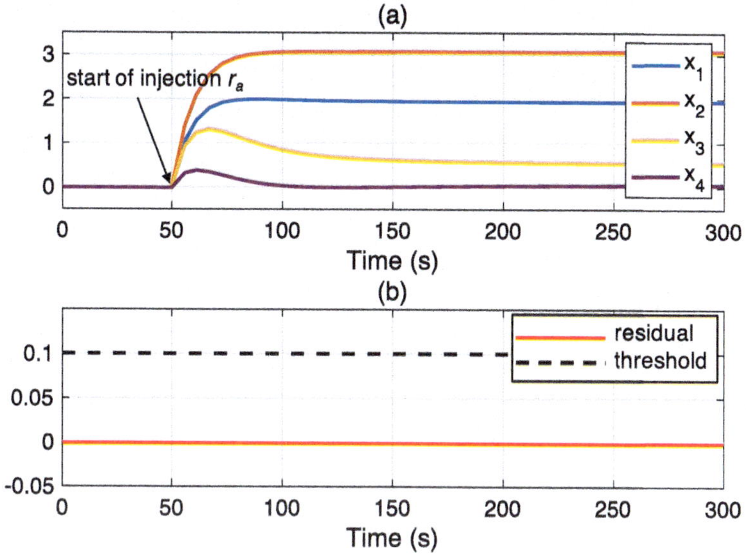

Fig. 4.2 Perfectly stealthy attack on Quadruple-Tank Process: **a** plant state x where x_i denotes the deviation of water level at the i-th tank from the stationary operating point, **b** residual $\|Wz\|$ and threshold $\gamma = 0.1$

It is assumed that the adversary starts launching system integrity attack by injecting exogenous signal r_a in (4.5) from time $t_0 = 50$ to manipulate the dynamic response of the physical plant. As can be observed from Fig. 4.2a, the adversary is able to make the state of the physical system deviate from the desired one. Furthermore, the attack cannot be detected by detector in (4.4) as the measurement residual between the measurement and observer output remains zero as shown in Fig. 4.2b.

4.2.2 Detecting Stealthy Attack

As demonstrated in the previous subsection, standard attack detection algorithm is not sufficient to detect an attack launched by intelligent adversary. To overcome this problem, the attack detection (4.4) will be enhanced by embedding nonlinear encoding/decoding components of chaotic signals as illustrated in Fig. 4.3.

Specifically, at the control center side, the control input which will be transmitted via the communication network is encoded by a nonlinear function generated by a time-varying signal. The compromised observed-based state feedback control with the encoding component can be written as

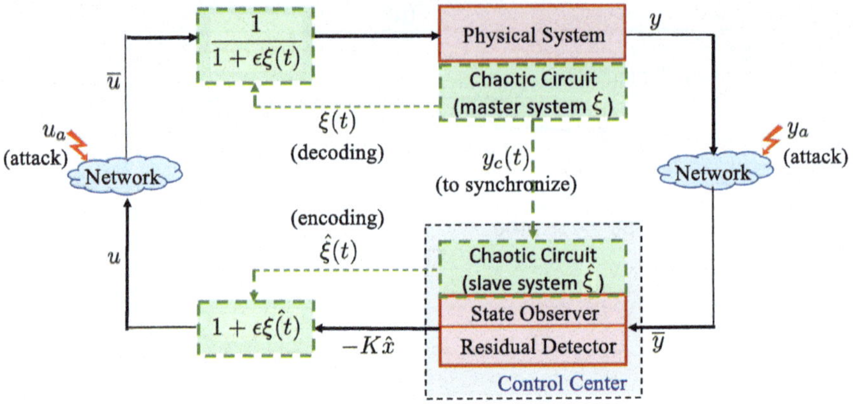

Fig. 4.3 Protection of networked control system against perfectly stealthy system integrity attacks by embedding nonlinear encoding/decoding components

$$\begin{aligned} \dot{\hat{x}} &= A\hat{x} + \frac{1}{1+\varepsilon\hat{\xi}}Bu + L(\overline{y} - \hat{y}), \\ u &= -[1+\varepsilon\hat{\xi}]K\hat{x}, \quad \hat{y} = C\hat{x}, \quad \overline{y} = y + y_a, \end{aligned} \tag{4.6}$$

where $\hat{\xi} \in \mathbb{R}$ denotes the *dynamic encoding signal* and $\varepsilon > 0$ is a scalar value. At the plant side, the encoded control input signal is decoded using another nonlinear function. Dynamics of the compromised physical plant with the decoding component is given by

$$\dot{x} = Ax + \frac{1}{1+\varepsilon\xi}B\overline{u}, \quad y = Cx, \quad \overline{u} = u + u_a, \tag{4.7}$$

where $\xi \in \mathbb{R}$ denotes the *dynamic decoding signal* and the scalar constant ε is chosen to ensure $\varepsilon|\xi| \leq 0.5$ and $\varepsilon|\hat{\xi}| \leq 0.5$. Intuitively speaking, by adding the nonlinear encoding/decoding components, the NCS structure (4.6), (4.7) does not hold the linearity property. As a result, it will not be possible for the adversary to launch perfectly stealthy attack using only information of A, B, and C. The encoding/decoding signals are generated such that the following conditions are satisfied: (i) the nominal performance of the NCS in the absence of attacks is preserved; (ii) the robustness of NCS against attacks is guaranteed. First, it can be observed from Fig. 4.3 that if both the decoding and encoding signals are synchronized, i.e., $\lim_{t\to\infty} |\xi(t) - \hat{\xi}(t)| = 0$, the NCS with encoding/decoding components (4.6), (4.7) will then recover its nominal performance. Hence, an additional *scalar* output signal y_c called the *synchronization signal* is introduced to guarantee synchronization between the encoding and decoding signals. The synchronization signal is transmitted from the plant side to the control center side, as shown in Fig. 4.3, so that the encoding signal at the control center side will synchronize to the decoding signal at the plant side. Due to this configuration, the physical plant is also called *master* system while the control center is called *slave* system.

The encoding/decoding signals can be generated by any types of chaotic circuits (oscillators), as long as they can be synchronized using an output feedback, which will make it very difficult for the adversary to imitate. One possible choice of such chaotic circuits is Chua's circuit [24, 25]. Specifically, the master system runs the following dynamics

$$\begin{aligned} C_1\dot{v}_1 &= \frac{1}{R_1}(v_2 - v_1) - g(v_1), \\ C_2\dot{v}_2 &= \frac{1}{R_1}(v_1 - v_2) + I, \\ L_1\dot{I} &= -v_2 - R_2 I, \quad y_c = v_1, \end{aligned} \tag{4.8}$$

where parameters L_1 is an inductor, R_1, R_2 are resistors, C_1, C_2 are capacitors, $v_1, v_2 \in \mathbb{R}$ denote the voltages, $I \in \mathbb{R}$ is the current, and $g(\cdot)$ is a nonlinear function defined as

$$g(v_1) = \begin{cases} \overline{d}v_1 + (\overline{d} - \underline{d})E & \text{if } v_1 \leq E, \\ \overline{d}v_1 > \gamma & \text{if } |v_1| < E, \\ \overline{d}v_1 + (\underline{d} - \overline{d})E & \text{if } v_1 \geq E \end{cases}$$

with $\underline{d} < -1/(R_1 + R_2) < \overline{d} < 0$ and $E > 0$ are constants. Similarly, the slave system runs the following dynamics using the signal received from the master system

$$\begin{aligned} C_1\dot{\hat{v}}_1 &= \frac{1}{R_1}(\hat{v}_2 - \hat{v}_1) - g(\hat{v}_1) + l_c(y_c - \hat{y}_c), \\ C_2\dot{\hat{v}}_2 &= \frac{1}{R_1}(\hat{v}_1 - \hat{v}_2) + \hat{I}, \\ L_1\dot{\hat{I}} &= -\hat{v}_2 - R_2\hat{I}, \quad \hat{y}_c = \hat{v}_1, \end{aligned} \tag{4.9}$$

where $l_c > 0$ is the coupling gain, $\hat{v}_1, \hat{v}_2, \hat{I}$ are the estimates of v_1, v_2, I, respectively. The parameters, initial conditions, and coupling gain l_c are chosen such that all the master system's variables v_1, v_2, I are both chaotic and uniformly bounded and further the errors $v_1 - \hat{v}_1, v_2 - \hat{v}_2, I - \hat{I}$ are exponentially stable, i.e., both chaotic oscillators are synchronized.

As can be observed, outputs of the chaotic circuits cannot be decoded without the full knowledge of its specific class and parameters together with real-time information of such nonlinear signal, which makes it very hard for the adversary to imitate the plant behavior properly. In addition, the synchronization signal y_c can be transmitted via a more secure communication channel, separated from the communication network used to transmit the control and output vectors, to further prevent any eavesdropping. In contrast to physical plant's control and output signals of higher dimensions and at different locations, securing a scalar signal such as the synchronization signal is much easier to accomplish.

After ensuring that, both the chaotic circuits can be synchronized, the next step is to choose both the encoding and decoding signals in (4.6), (4.7). For example, both $\xi, \hat{\xi}$ can be chosen as

$$\xi = \alpha_1 v_1 + \alpha_2 v_2 + \alpha_3 I, \quad \hat{\xi} = \alpha_1 \hat{v}_1 + \alpha_2 \hat{v}_2 + \alpha_3 \hat{I} \tag{4.10}$$

with $\alpha_i \geq 0$ and $\sum_i \alpha_i = 1$. For the parameters and coupling gain chosen to ensure synchronization between the master and slave systems, it can be shown that the NCS (4.6), (4.7) with encoding/decoding components generated by chaotic circuits (4.8), (4.9) is global asymptotically stable in the absence of attacks and also input-to-state stable with respect to attack vector u_a, y_a. In addition, after the two oscillators achieving synchronization the perturbed measurement residual $(z - z_n)$ can be calculated as

$$z - z_n = -C \int_{t_0}^{t} \frac{\varepsilon\xi(\eta)}{1 + \varepsilon\xi(\eta)} e^{(A-LC)(t-\eta)} B u_a(\eta) d\eta,$$

which equals to zero (i.e., $z = z_n$) only if there is no attack (or any attack affecting the system), i.e., $u_a = 0$ (or $Bu_a = 0$). Hence, the perfectly stealthy attack can be detected using residual-based attack detector (4.4). Note that in practice, the control center operator needs to choose ε to exceed the detector threshold γ since the magnitude of $z - z_n$ increases as ε increases. Furthermore, since it is, in general, difficult to generate attack vectors u_a, y_a to compensate the effect of cyberattacks, the proposed NCS structure can also be utilized to detect other types of stealthy attacks.

Let us now return to the previous example of Quadruple-Tank process. As demonstrated in Sect. 4.2.1, an intelligent adversary is able to launch a perfectly stealthy attack in order to manipulate dynamic response of the physical plant. Figure 4.4 shows simulation results for the case of applying the proposed attack-aware NCS structure with embedded encoding/decoding components. As can be observed from Fig. 4.4b, as the attack being launched, the residual shows the oscillating behavior and as a result, the residual detector is triggered. Hence, it can be confirmed that the perfectly stealthy system integrity attack can be detected. The proposed attack-aware NCS structure is also scalable for networked physical systems as the synchronization signal can be multi-cast from the master system to all physical systems equipped with their own slaves.

4.3 Resilient and Safe Control at the Edge

Cooperative control is one of control design tools that has shown a great promise in optimizing and controlling thousands to millions of devices at the edge of networked control systems [13, 30]. Briefly speaking, the objective of cooperative control is to steer the entity of interest (i.e., physical variable) of each individual system, denoted

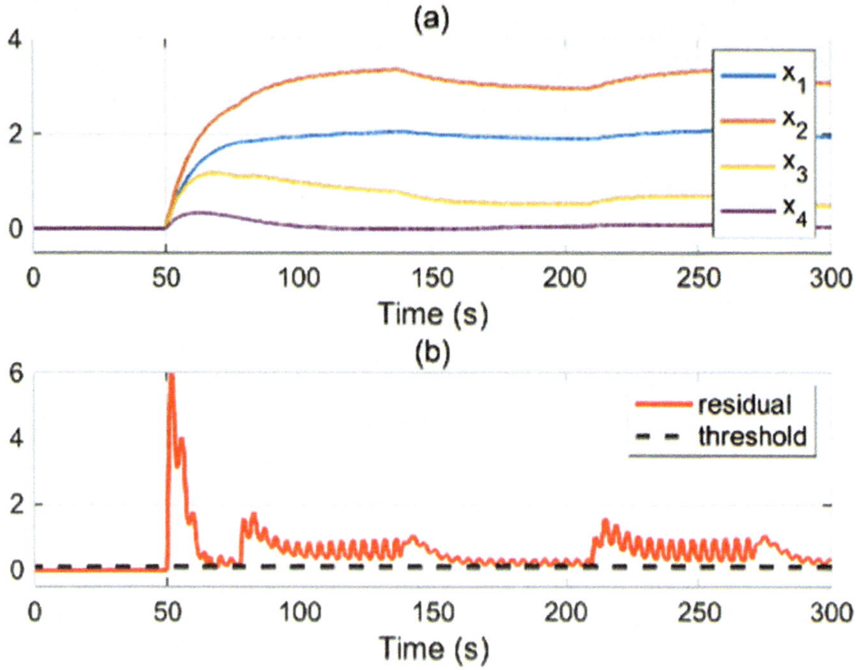

Fig. 4.4 Detection of perfectly stealthy attack by embedding nonlinear encoding/decoding components: **a** plant state x where x_i denotes the deviation of water level at the i-th tank from the stationary operating point, **b** residual $\|Wz\|$ and threshold $\gamma = 0.1$

by $x_i \in \mathbb{R}$, to achieve a non-trivial consensus using only local information. In other words, for an NCS consisting of n individual subsystems, we have

$$\lim_{t\to\infty} x_i(t) = c^* \text{ for all individual system } i, \tag{4.11}$$

or

$$\lim_{t\to\infty} x(t) = x^* = c^*\mathbf{1}, \tag{4.12}$$

where $x = [x_1, x_2, \ldots, x_n]^T$ and $c^* \in \mathbb{R}$ denotes the nominal operating condition of the overall system. The entity of interest x_i can vary depending on the applications. For example, in optimization and control of power system, x_i can represent the utilization ratio of active power [35] or reactive power [23] produced by the individual distributed generation. Cooperative control which achieves the objective (4.12) in general is given by the following dynamics

$$\dot{x}_i = \sum_{j=1}^{n} s_{ij}(x_j - x_i), \tag{4.13}$$

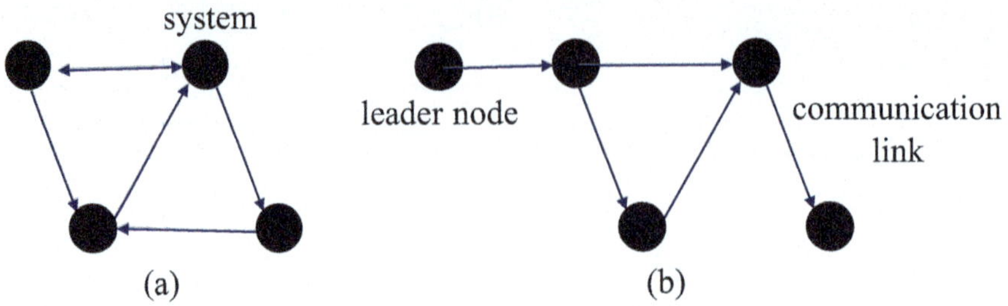

Fig. 4.5 **a** Leaderless, **b** leader-following information structure (communication network topology) for cooperative control algorithm

where $s_{ij} = 1$ if the i-th system can receive information from system j via the communication network (i.e., system j is a neighbor of system i) and $s_{ij} = 0$ otherwise. The cooperative control (4.13) can be written in a compact form as

$$\dot{x} = -L_s x, \tag{4.14}$$

where L_s is also called a Laplacian matrix which satisfies $L_s \mathbf{1} = 0$. The nominal operating condition c^* in (4.12) is calculated depending on the information structure (i.e., communication network topology) of the cooperative control algorithms which can be summarized as follows:

1. *Leaderless* cooperative control whose information structure is shown in Fig. 4.5a. Specifically, the communication network structure is given by a strongly connected directed graph, that is, every node/system can be reached from any other nodes by following a set of directed edges (communication links) as depicted in Fig. 4.5a. The nominal operating condition of leaderless cooperative control depends on initial conditions of all the systems in the network. Specifically, it is given by $c^* = \nu_1^T x(0)$ where ν_1 denotes the left eigenvector corresponding to zero eigenvalue of the Laplacian matrix L_s and $x(0)$ is the initial condition of the state $x(t)$. Leaderless cooperative control has been used in different applications, for example, to distributively regulate the frequency in power network as discussed in [38].
2. *Leader-following* cooperative control whose communication network topology is illustrated in Fig. 4.5b. Here the leader node (aggregator/control center/higher level control) provides a reference value, denoted by x_r, that needs to be tracked by all the systems in the network. For leader-following case, cooperative control in (4.14) can be written as

$$\dot{x} = Ax + Bx_r, \tag{4.15}$$

where matrix A is Hurwitz and vector $B \in \mathbb{R}^n$ with its i-th element equals to one if node i can receive information from the leader node and zero otherwise. The consensus (4.12) is guaranteed given that there exists a sequence of edges (communication links) from the leader node to every other nodes in the network as depicted in Fig. 4.5b. In addition, the operating condition x^* is given by $x^* = x_r\mathbf{1}$. Leader-following cooperative control has been utilized to distributively regulate the power output of multiple photovoltaic generators in a distribution network [35] and also for real-time scheduling of electric vehicles' charging at a highway [11].

3. The last one is the combination of cooperative control (4.13) and distributed optimization algorithm designed to solve the following optimization problem

$$\begin{aligned} \underset{x_1,\ldots,x_n}{\text{minimize}} \quad & \sum_{i=1}^{n} f_i(y_i) \\ \text{subject to} \quad & x_1 = \cdots = x_n, \\ & h(y_1,\ldots,y_n,x_1,\ldots,x_n) = 0, \end{aligned} \tag{4.16}$$

where function $f_i(y_i)$ is strictly convex w.r.t. decision variable x_i. Cooperative control algorithm to solve optimization problem (4.16) is given by [37]

$$\dot{x}_i = \sum_{j=1}^{n} s_{ij}(x_j - x_i) - k_i g_i(x_i), \tag{4.17}$$

where $k_i > 0$ is a step size gain and g_i is the subgradient of f_i w.r.t. x_i. Cooperative control (4.17) can also be written in a compact form as

$$\dot{x} = -L_s x - \overline{K} g(x), \tag{4.18}$$

where $\overline{K} = diag\{k_i\}$, $g(x) = [g_1(x_1), \ldots, g_n(x_n)]^T$ and the communication network topology is given by a connected bidirectional graph. Furthermore, the operating condition x^* is given by the solution to optimization problem (4.16). This particular class of cooperative control algorithms has been used to control the reactive power of distributed generations in a distribution network so that the bus voltages are maintained within a certain limit [14, 23].

4.3.1 Cooperative Control and Intelligent Adversary

While the information and communication technologies (ICT) facilitates the implementation of cooperative control algorithms, it is known that ICT is vulnerable to cyber intrusions. The adversary may distort the communication channel by adding exogenous signals to modify the neighbors' information that a specific system receives. Therefore, the cooperative control (4.13) under potential attacks can be written as

$$\dot{x}_i = \sum_{j=1}^{n} s_{ij}(\tilde{x}_{i,j} - \tilde{x}_{i,i}), \tag{4.19}$$

where

$$\tilde{x}_{i,j} = x_j + \delta_{i,j}, \quad j \in \mathcal{N}_i \cup \{i\}.$$

Here, $\mathcal{N}_i$ is the set of neighbors of node j, i.e., node $j \in \mathcal{N}_i$ if and only if $s_{ij} = 1$ and $\delta_{i,j}$ denotes the injection inserted by the attacker. The attacker aims at destabilizing the overall system or steering the system outside its safety operational constraint without being noticed. In the worst case, the adversary may gain the full knowledge of the network (i.e., matrix L_s), and he/she may attempt to compromise as many as communications links as possible by false data injection. It will be demonstrated in Sect. 4.3.3 that the adversary could make the system violate its operational constraint by compromising a small fraction of communication links with bounded injection. The reason that the adversary does not need to have much (or the full) knowledge of the system is because, unless there is a designated leader, a consensus-based cooperative system treats all the entities equally and let all the information propagate throughout the system. Since it is not possible to secure all the communication channels, in the following, a resilient cooperative control algorithm is presented to ensure safe operation of the NCS against unknown attacks.

4.3.2 *Resilient and Safe Cooperative Control*

The objective is to ensure resilient and safe operation of cooperative controls described previously against unknown attacks. In other words, in addition to ensuring the physical variable x to be bounded, we also aim at maintaining the steady state of x within a safe region around the nominal operating condition, that is

$$\|x(t) - x^*\| \leq \varepsilon \tag{4.20}$$

for a large t where ε is a pre-defined threshold value. The definition of safety in (4.20) is motivated by a variety of stability problems in power system. For example, it is important to maintain the frequency or voltages in power grid tightly around the nominal operating value, which can be formulated as in (4.20).

Design of resilient and safe cooperative control consists of two main steps. In the first step, each subsystem in the network checks and removes the excessively large values of information that it receives from its neighbors. Mathematically, this step can be written as

$$s_{ij} = \begin{cases} 1 & \text{if } \|\tilde{x}_{i,j}\| < \sigma, \\ 0 & \text{otherwise,} \end{cases} \tag{4.21}$$

where σ is a threshold chosen by the designer. The threshold σ is chosen depending on the operational range of the physical variables to be controlled. Threshold-based strategy in (4.21) can also be utilized to isolate faulty system in the network. After applying (4.21), the leaderless cooperative control under potential attacks can be written as

$$\dot{x} = -L_s(x - d), \tag{4.22}$$

where $d_i = \sum_{j=1}^{n} \delta_{i,j}$. Similarly, we can write, respectively, the leader-following cooperative control and distributed algorithm for solving optimization problem (4.16) under potential attacks as

$$\dot{x} = Ax + Bx_r + d, \tag{4.23}$$

and

$$\dot{x} = -L_s(x - d) - \overline{K}g(x). \tag{4.24}$$

Due to (4.21), it can be observed that the injection $\|d\|$ is bounded. Note that if there exists faulty systems isolated from the network as a result of (4.21), then the networked control systems (4.22), (4.23) and (4.24) will be composed of a smaller number of subsystems than the original system, called healthy subsystems and the matrix L_s or A, B represent the communication network topology between the healthy subsystems/nodes. Hence, the above framework can be utilized to address both cyber-attacks and also faulty subsystems.

The injection d may be chosen by the adversary to steer the steady state of the networked control system away from the nominal operating condition x^*, i.e., violating (4.20). In order to overcome this issue, in the second step, a resilient and safe cooperative control will be designed by introducing a virtual system with the same number of nodes as the physical system and interconnected with the cooperative system as illustrated in Fig. 4.6. The virtual system in principle acts as an anchor which will maintain the networked control systems to operate around its optimal operating point under unknown attacks. The cooperative controls (4.22), (4.23), (4.24) interconnected with the virtual system are, respectively, given by

$$\text{leaderless} \begin{cases} \dot{x} = -L_s(x - d) + \beta\Omega z, \\ \dot{z} = -L_h z - \beta\Upsilon x, \end{cases} \tag{4.25}$$

$$\text{leader-following} \begin{cases} \dot{x} = Ax + \beta\Omega z + Bx_r + d, \\ \dot{z} = A_h z - \beta\Upsilon x + \beta B_h x_r, \end{cases} \tag{4.26}$$

and

$$\text{distributed optimization} \begin{cases} \dot{x} = -L_s(x - d) - \beta\overline{K}g + \beta\Omega z, \\ \dot{z} = -L_h z - \beta\Upsilon x. \end{cases} \tag{4.27}$$

Here, z is the state of the virtual system, scalar $\beta > 0$ is the control gain which adjusts the strength of interconnection between the cooperative and virtual systems, matrices A_h, L_h denote the "internal" dynamics of the virtual system, and Ω, Υ, B_h are the interconnection matrices. The virtual systems (i.e., matrices A_h, L_h) together with its

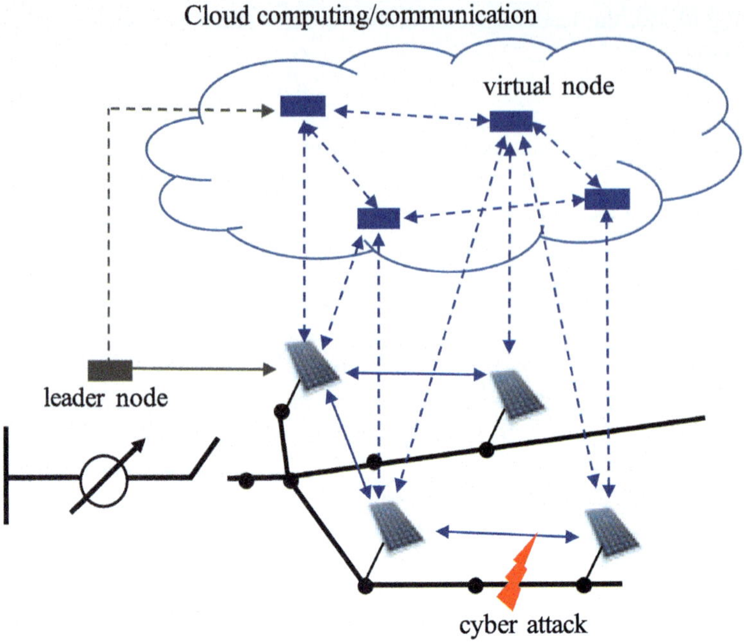

Fig. 4.6 Resilient design of cooperative control: Interconnection of the cooperative system and virtual system. The dashed-line represent information flow for robustification of cooperative system

interconnection (i.e., matrices Ω, Υ, B_h) and control gain β need to be designed such that: (i) its interconnection with the networked systems does not impact convergence of the cooperative control; (ii) the robustification strategy is automatically activated when attacks appear anywhere in the system; (iii) the virtual nodes maintain stability of the networked control system under bounded attack signals, i.e., mitigation measures embedded. These properties are highly desirable in designing resilient control algorithms since in practice the attacks cannot be foreseen in advance.

The design of virtual systems in (4.25), (4.26) and (4.27) can be summarized as follows.

1. For the resilient leaderless cooperative control (4.25), matrix L_h is chosen to be a Laplacian matrix corresponding to arbitrary strongly connected directed graph. Furthermore, we set $\Omega = L_s$ and matrix Υ is chosen as $\Upsilon = \Gamma_h^{-1} L_s^T \Gamma_s$ where $\Gamma_s = diag\{\nu_1\}$ and $\Gamma_h = diag\{\nu_{h1}\}$ with $\nu_{h1}^T L_h = 0$.
2. For the resilient leader-following cooperative control (4.26), matrix A_h is chosen to be any sparse Hurwitz matrix. Furthermore, Ω is chosen to be any invertible sparse matrix and $\Upsilon = P_h^{-1} \Omega^T P$ where matrices $P, P_h > 0$ satisfy $A^T P + PA < 0$ and $A_h^T P_h + P_h A_h < 0$, respectively. Vector B_h is then calculated as $B_h = \Upsilon \mathbf{1}$.

3. For resilient distributed optimization algorithm (4.27), we set matrices $L_h = \Omega = \Upsilon = M\Lambda^{\frac{1}{2}}M^T$ where $\Lambda = diag\{0, \lambda_2(L_s), \ldots, \lambda_n(L_s)\}$ and $M = [\nu_1, \ldots, \nu_n]$ with ν_i is the eigenvector of symmetric matrix L_s associated with the eigenvalue $\lambda_i(L_s)$.

In addition, using Lyapunov stability analysis, it can be shown that for a sufficiently large value of β and the above choices of interconnection matrices, the state x converges to the operating point around x^*. Specifically, the steady state of physical variable x for leaderless and leader-following cooperative control algorithms can be explicitly calculated as

$$\begin{aligned} \text{leaderless: } & \lim_{t\to\infty} x(t) = x^* + (I + \beta^2 M_g)^{-1}[c_1\mathbf{1} + \beta c_2\mathbf{1} + d^e], \\ \text{leader-following: } & \lim_{t\to\infty} x(t) = x^* - (A + \beta^2 \Omega A_h^{-1} P_h^{-1} \Omega^T P)^{-1} d^e, \end{aligned} \tag{4.28}$$

where c_1, c_2 are some constants which depend upon initial conditions $x(0), z(0)$, and $d(0)$, matrix M_g satisfies $\Upsilon = L_h M_g$. Note that $d(t)$ can be decomposed into $d(t) = d^e + \tilde{d}(t)$. Hence, from (4.28) minimum value of the control gain β which ensures the perturbed NCS to operate within a distance ε from its nominal operating condition x^* can be calculated. In practice, the designer can use the upper bound of d^e to calculate the minimum gain β from (4.28), for example, by simply setting $d^e = \sigma$ where σ is defined in (4.21). For distributed algorithm (4.27), we can write

$$L_h x^e = (L_h + \beta^2 I)^{-1}[L_s d^e - \beta \overline{K} g(x^e)],$$

where $x^e = \lim_{t\to\infty} x(t)$. It can be observed that $x^e \to c\mathbf{1}$ as control gain β increases. Furthermore, noting that function f_i is strictly convex, we have $c \to c^*$ for sufficiently large $\beta > 0$. Hence, the control gain β can be adaptively adjusted so that x_i approximately reaches consensus, i.e., $\|x - c\mathbf{1}\| < \varepsilon$ and since the subgradient $g(x^e)$ is bounded, it can be observed that x converges to a point around $c^*\mathbf{1}$ accordingly.

The virtual system can be realized by taking advantage of the cloud computing/communication in combination with software-defined networking (SDN) [20] which has been shown to be a promising architecture for promoting distributed decision-making in cyber-physical systems [7]. Briefly speaking, SDN provides a flexibility to direct traffic in a network as it separates the part of the networking infrastructure that decides where information is being sent from the part where the data actually moves, and allows the decision-making part to happen in the software application. In addition, as its name suggests the virtual node is not a physical node and its state has no physical meaning which makes it less observable to the adversary. While the addition of such a virtual system incurs an additional cost of increased communication, such a burden is manageable and the corresponding computation is minimal since both the communication and computation are performed distributively and using one of the standard network technologies. The framework has been recently utilized to deal with partially unknown nonlinear systems [15] and also multiple types of actuator attacks [16].

4.3.3 An Illustrative Example

The resilient and safe cooperative control described in the previous subsection is applied to voltage control problem in a distribution network with high penetration of renewable energy sources [14]. High penetration of renewable energy sources or distributed generations (DGs) in distribution network may lead to local congestion, e.g., over-voltage problem. One possible approach to overcome this issue is by distributively controlling the reactive power of DGs such as the voltage deviation of all DG buses satisfy the operational constraint of

$$|1 - V_i| \leq 0.05 p.u.$$

where V_i denotes the voltage of the ith DG. The problem can then be formulated as optimization problem in the form of (4.16) given by

$$\begin{aligned} \underset{x_1,\ldots,x_n}{\text{minimize}} \quad & \sum_{i=1}^{n}(V_i - V_i^*)^2 \\ \text{subject to} \quad & x_1 = \cdots = x_n, \\ & \text{AC power flow}, \end{aligned} \tag{4.29}$$

where x_i denotes the utilization ratio of reactive power of DG i. The first constraint means that all DGs contribute equally to the voltage regulation while the second constraint physically couples the decision variable x_i with the voltage V_i. The values $V_i^* \in (0.95, 1.05)$ denote the operating condition of the system voltages in order to ensure power quality and are calculated at a higher level control authority by solving a distributed optimal power flow [36]. Given that the total maximum available reactive power is sufficient to regulate the voltages, then there exists a consensus solution x^* to optimization (4.29). The solution can be calculated in a distributed manner using distributed algorithm (4.18), where the communication graph is assumed to be connected and undirected.

Cooperative control (4.18) is implemented in IEEE 8500-node system consisting of seven photovoltaics (PVs) as shown in Fig. 4.7 and communication network topology of the PVs is given in Fig. 4.8.

First, consider the nominal operation under cooperative control (4.18) in the absence of attacks. As can be observed from Fig. 4.9a and b, cooperative control (4.18) is able to maintain the system voltages within the limit while ensuring all PVs contribute equally to the voltage regulation. Next, assume that the information sent from PV-6 is being compromised. Simulation results for cooperative control under attack (4.24) are shown in Fig. 4.10. As can be observed from Fig. 4.10a, even though the states x are still bounded, the adversary is able to make the system voltages violate the operational constraint, i.e., yielding an over-voltage problem, by injecting bounded signal into the communication link. In addition, we can see from Fig. 4.10b that the reactive power utilization ratios also fail to reach a consensus.

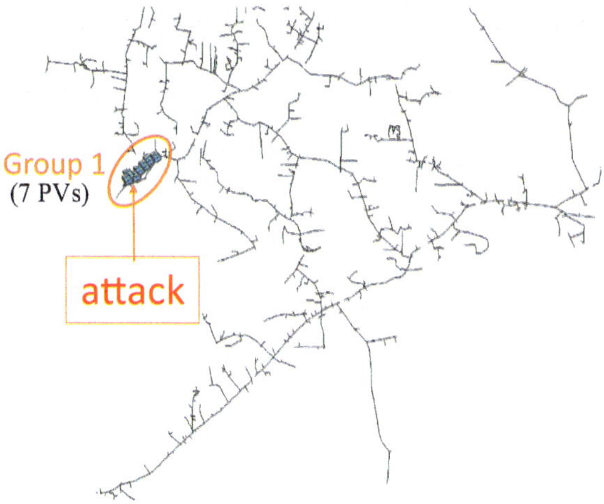

Fig. 4.7 IEEE 8500-node system

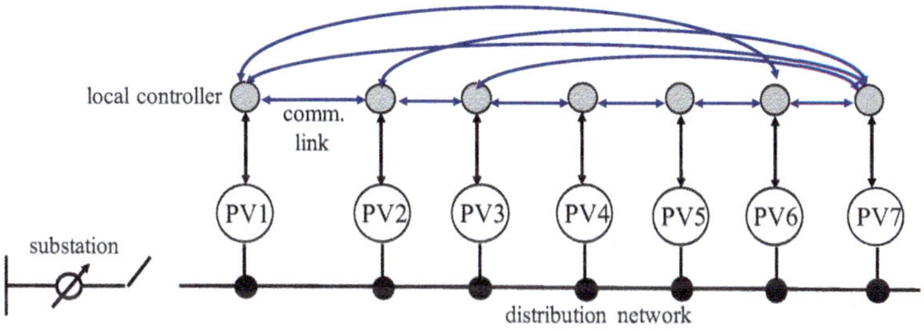

Fig. 4.8 Communication network topology for cooperative control

Finally, in order to maintain resilient and safe operation of the system voltages, we apply resilient cooperative control (4.27) under attacks whose results are shown in Fig. 4.11. As can be observed, the system voltages are regulated within the operational constraint and the reactive power utilization ratios are close to the nominal operating condition x^*.

4.4 Conclusions

The chapter presents resilient hierarchical networked control systems. First, a variant of dynamic watermarking strategies is presented by embedding encoding/decoding components of chaotic signals into the NCS for secure control for critical locations where the measurement/control signals are transmitted to/from the control center via a communication network. Next, resilient cooperative control algorithms by introducing a virtual system which acts as an anchor are discussed to ensure safe operation

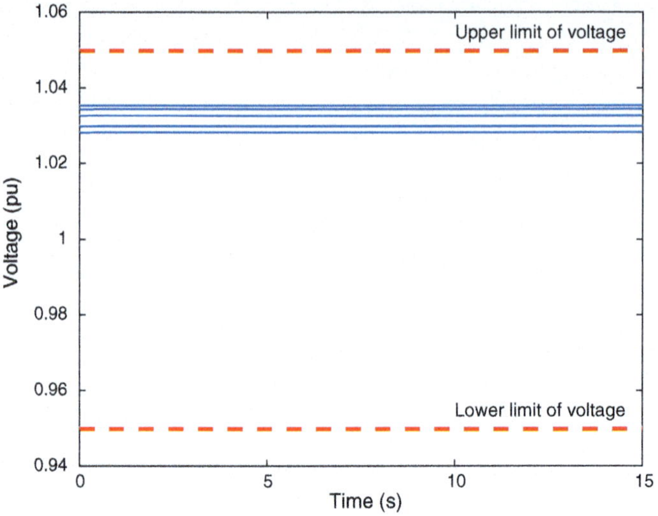

(a) Voltage profile of PVs in the absence of attacks

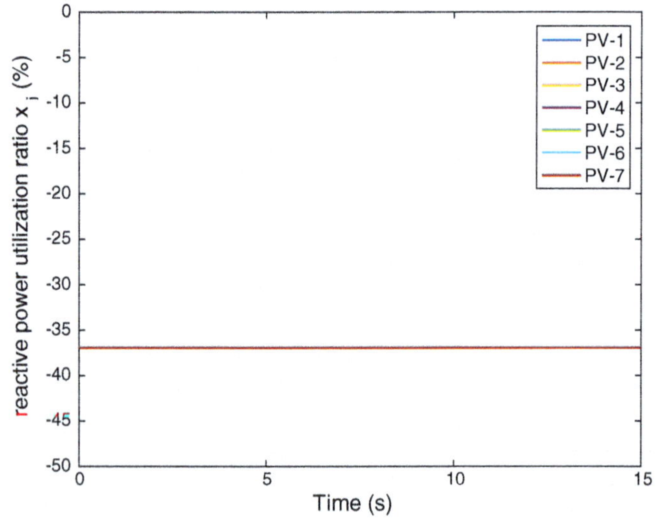

(b) Reactive power utilization ratio of PVs in the absence of attacks

Fig. 4.9 Cooperative control (4.18) in the absence of attacks

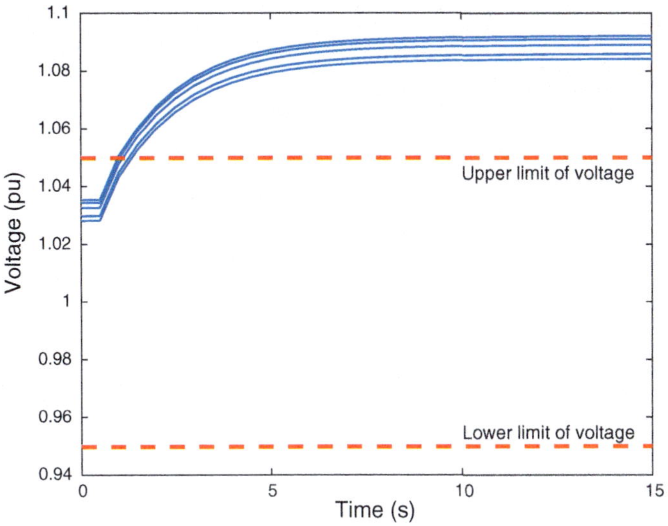

(a) Voltage profile of PVs under attacks

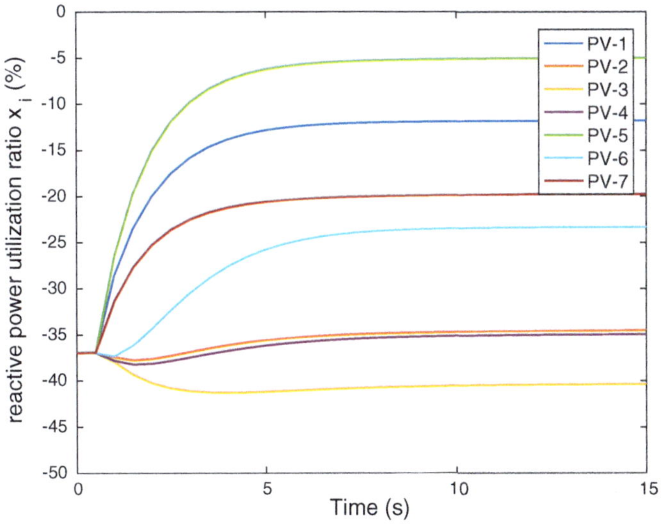

(b) Reactive power utilization ratio of PVs under attacks

Fig. 4.10 Cooperative control under attack (4.24). System voltages violate the operational constraint

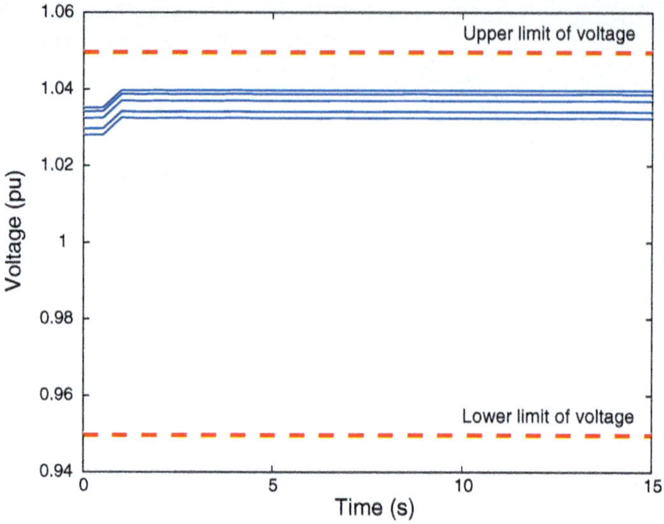

(a) Voltage profile of PVs under attacks

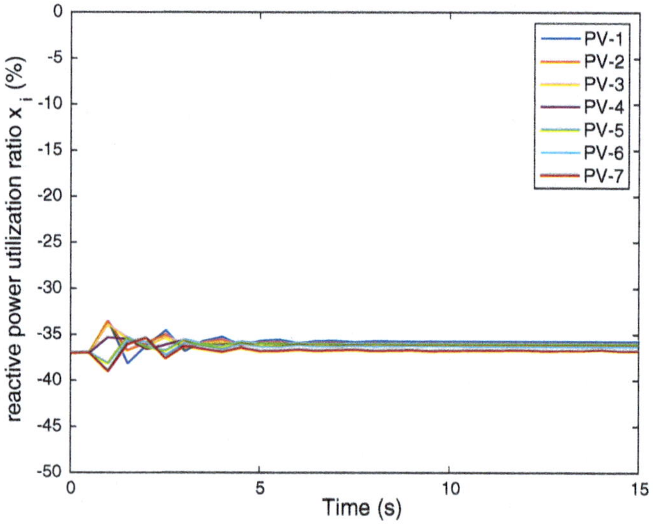

(b) Reactive power utilization ratio of PVs under attacks

Fig. 4.11 Resilient and safe cooperative control against attack (4.27). System voltages can be maintained within the operational constraint

at the edge of the NCS which consists of a large number of control devices. The proposed control algorithms can be implemented for using the state-of-the-art networking technologies such as cloud computing/communication and software-defined networking. The performance of resilient control strategies are demonstrated using several numerical examples.

Acknowledgements The work of A. Gusrialdi was supported by the Academy of Finland under Academy Project decision number 330073. The work of Z. Qu is supported in part by US National Science Foundation under grant ECCS-1308928, by US Department of Energy's awards DE-EE0009152, DE-EE0009028, DE-EE0007998, DE-EE0007327, DE-EE0006340, and DE-EE0009339.

References

1. Abhinav, S., Modares, H., Lewis, F.L., Davoudi, A.: Resilient cooperative control of dc microgrids. IEEE Trans. Smart Grid **10**(1), 1083–1085 (2019)
2. Blume, S..W.: High Voltage Protection for Telecommunications. IEEE Press Series on Power and Energy Systems, vol. 44. Wiley (2011)
3. Cameron, C., Patsios, C., Taylor, P., Pourmirza, Z.: Using self-organizing architectures to mitigate the impacts of denial-of-service attacks on voltage control schemes. IEEE Trans. Smart Grid **10**(3), 3010–3019 (2019)
4. Chakrabortty, A., Khargonekar, P.P.: Introduction to wide-area control of power systems. In: Proceedings of American Control Conference, pp. 6758–6770 (2013)
5. Dibaji, S.M., Pirani, M., Flamholz, D.B., Annaswamy, A.M., Johansson, K.H., Chakrabortty, A.: A systems and control perspective of cps security. Ann. Rev. Control **47**, 394–411 (2019)
6. Frank, P.M.: Fault diagnosis in dynamic systems using analytical and knowledge-based redundancy: a survey and some new results. Automatica **26**(3), 459–474 (1990)
7. Freris, N.M., et al.: A software defined architecture for cyberphysical systems. In: Proceedings of 4th International Conference on Software Defined Systems, pp. 54–60 (2017)
8. Gusrialdi, A., Qu, Z.: Smart grid security: attacks and defenses. In: Stoustrup, J., Annaswamy, A., Chakrabortty, A., Qu, Z. (eds.) Smart Grid Control: An Overview and Research Opportunities, pp. 199–223. Springer (2018)
9. Gusrialdi, A., Qu, Z., Simaan, M.: Robust design of cooperative systems against attacks. In: Proceedings of American Control Conference, pp. 1456–1462 (2014)
10. Gusrialdi, A., Qu, Z., Simaan, M.A.: Game theoretical designs of resilient cooperative systems. In: Proceedings of European Control Conference, pp. 1705–1711 (2015)
11. Gusrialdi, A., Qu, Z., Simaan, M.A.: Distributed scheduling and cooperative control for charging of electric vehicles at highway service stations. IEEE Trans. Intell. Transp. Syst. **18**(10), 2713–2727 (2017)
12. Gusrialdi, A., Qu, Z., Simaan, M.A.: Competitive interaction design of cooperative systems against attacks. IEEE Trans. Autom. Control **63**(9), 3159–3166 (2018)
13. Gusrialdi, A., Xu, Y., Qu, Z., Simaan, M.A.: A real-time big-data control-theoretical framework for cyber-physical human systems. In: Blondin, M.J., Pardalos, P.M., Sanchis, J.S. (eds.) Computational Intelligence and Optimization Methods for Control Engineering, pp. 149–172. Springer (2019)
14. Gusrialdi, A., Xu, Y., Qu, Z., Simaan, M.A.: Resilient cooperative voltage control for distribution network with high penetration distributed energy resources. In: Proceedings of European Control Conference, pp. 1533–1539 (2020)

15. Huang, X., Dong, J.: ADP-based robust resilient control of partially unknown nonlinear systems via cooperative interaction design. IEEE Trans. Syst. Man Cybern. Syst. (2020)
16. Huang, X., Dong, J.: A robust dynamic compensation approach for cyber-physical systems against multiple types of actuator attacks. Appl. Math. Comput. **380**, 125284 (2020)
17. Huang, Y., Tang, J., Cheng, Y., Li, H., Campbell, K.A., Han, Z.: Real-time detection of false data injection in smart grid networks: an adaptive CUSUM method and analysis. IEEE Syst. J. **10**(2), 532–543 (2014)
18. Johansson, K.H.: The quadruple-tank process: a multivariable laboratory process with an adjustable zero. IEEE Trans. Control Syst. Technol. **8**(3), 456–465 (2000)
19. Joo, Y., Qu, Z., Namerikawa, T.: Resilient control of cyber-physical system using nonlinear encoding signal against system integrity attacks. IEEE Trans. Autom. Control **66**(9), 4334–4341 (2021)
20. Kreutz, D., Ramos, F.M., Verissimo, P., Rothenberg, C.E., Azodolmolky, S., Uhlig, S.: Software-defined networking: a comprehensive survey. Proc. IEEE **103**(1), 14–76 (2015)
21. LeBlanc, H.J., Zhang, H., Koutsoukos, X., Sundaram, S.: Resilient asymptotic consensus in robust networks. IEEE J. Sel. Areas Commun. **31**(4), 766–781 (2013)
22. Liu, X., Qian, C., Hatcher, W.G., Xu, H., Liao, W., Yu, W.: Secure Internet of Things (IoT)-based smart-world critical infrastructures: survey, case study and research opportunities. IEEE Access **7**, 79523–79544 (2019)
23. Maknouninejad, A., Qu, Z.: Realizing unified microgrid voltage profile and loss minimization: a cooperative distributed optimization and control approach. IEEE Trans. Smart Grid **5**(4), 1621–1630 (2014)
24. Matsumoto, T.: A chaotic attractor from Chua's circuit. IEEE Trans. Circuits Syst. **31**(12), 1055–1058 (1984)
25. Matsumoto, T., Chua, L., Komuro, M.: The double scroll. IEEE Trans. Circuits Syst. **32**(8), 797–818 (1985)
26. Mo, Y., Weerakkody, S., Sinopoli, B.: Physical authentication of control systems: designing watermarked control inputs to detect counterfeit sensor outputs. IEEE Control Syst. Mag. **35**(1), 93–109 (2015)
27. Molzahn, D.K., Dörfler, F., Sandberg, H., Low, S.H., Chakrabarti, S., Baldick, R., Lavaei, J.: A survey of distributed optimization and control algorithms for electric power systems. IEEE Trans. Smart Grid **8**(6), 2941–2962 (2017)
28. Pasqualetti, F., Bicchi, A., Bullo, F.: Consensus computation in unreliable networks: a system theoretic approach. IEEE Trans. Autom. Control **57**(1), 90–104 (2011)
29. Pultarova, T.: Cyber security - Ukraine grid hack is wake-up call for network operators [news briefing]. Eng. Technol. **11**(1), 12–13 (2016)
30. Qu, Z.: Cooperative Control of Dynamical Systems: Applications to Autonomous Vehicles. Springer (2009)
31. Satchidanandan, B., Kumar, P.R.: Dynamic watermarking: active defense of networked cyber-physical systems. Proc. IEEE **105**(2), 219–240 (2016)
32. Teixeira, A., Shames, I., Sandberg, H., Johansson, K.H.: Revealing stealthy attacks in control systems. In: Proceedings of 50th Annual Allerton Conference on Communication, Control, and Computing (Allerton), pp. 1806–1813 (2012)
33. Weerakkody, S., Sinopoli, B.: Detecting integrity attacks on control systems using a moving target approach. In: Proceedings of 54th IEEE Conference on Decision and Control (CDC), pp. 5820–5826 (2015)
34. Weerakkody, S., Sinopoli, B.: Challenges and opportunities: cyber-physical security in the smart grid. In: Stoustrup, J., Annaswamy, A., Chakrabortty, A., Qu, Z. (eds.) Smart Grid Control: An Overview and Research Opportunities, pp. 257–273. Springer (2019)
35. Xin, H., Qu, Z., Seuss, J., Maknouninejad, A.: A self-organizing strategy for power flow control of photovoltaic generators in a distribution network. IEEE Trans. Power Syst. **26**(3), 1462–1473 (2010)
36. Xu, Y., Sun, W., Qu, Z.: Renewable energy integration and system operation challenge: control and optimization of millions of devices. In: Jiang, H., Zhang, Y., Muljadi, E. (eds.) New Technologies for Power System Operation and Analysis. Elsevier (2020)

37. Yang, T., Yi, X., Wu, J., Yuan, Y., Wu, D., Meng, Z., Hong, Y., Wang, H., Lin, Z., Johansson, K.H.: A survey of distributed optimization. Ann. Rev. Control **47**, 278–305 (2019)
38. Zhao, C., Mallada, E., Dörfler, F.: Distributed frequency control for stability and economic dispatch in power networks. In: Proceedings of American Control Conference, pp. 2359–2364 (2015)
39. Zhuang, P., Deng, R., Liang, H.: False data injection attacks against state estimation in multi-phase and unbalanced smart distribution systems. IEEE Trans. Smart Grid **10**(6), 6000–6013 (2019)

Chapter 5
A Tutorial on Security and Privacy Challenges in CPS

Seyed Mehran Dibaji, Alefiya Hussain, and Hideaki Ishii

Abstract Cyber-Physical Systems (CPS) refer to the comprehensive integration of communication and control into physical systems. The communication network plays a crucial role in the stability of CPSs but it inadvertently exposes the control system to novel security and privacy threats. In this chapter, we carry out a survey of attacks that threaten the stability of control systems in CPSs along with a range of defenses. We organize the capabilities of CPSs into three domains: sensing mechanisms, communication infrastructure, and control strategies, and provide a mapping of deception, disruption, and disclosure attacks in each domain. We also discuss common approaches to designing resilient systems based on techniques drawn from game theory, machine learning, and control theory. We envision that as attacks increase in sophistication over the years, defense and resiliency techniques will also significantly evolve and expand in their capabilities. Hence, in this chapter, we lay out the foundational security and privacy threats and their defenses to illustrate the challenge of building resilient systems of the future.

5.1 Introduction

Over the past decade, integration of information, communication, and computation with physically engineered systems has made significant progress. The extent of the impact can be seen from the term Cyber-Physical Systems (CPS) now commonly

---

S. M. Dibaji (✉)
Department of Mechanical Engineering, Massachusetts Institute of Technology, Cambridge 02139, USA
e-mail: dibaji@mit.edu

A. Hussain
USC/Information Sciences Institute, Marina del Rey 90292, USA
e-mail: hussain@isi.edu

H. Ishii
Department of Computer Science, Tokyo Institute of Technology, Yokohama 226-8502, Japan
e-mail: ishii@c.titech.ac.jp

H. Ishii and Q. Zhu (eds.), *Security and Resilience of Control Systems*,
Lecture Notes in Control and Information Sciences 489,
https://doi.org/10.1007/978-3-030-83236-0_5

used in the different communities in the industry, government, and research. In particular, such advances have enabled instrumentation, monitoring, and automation to be enhanced for the objective of improving sustainability, efficiency, and resiliency. New technologies have reached different sectors of critical infrastructure such as energy, transportation, water, agriculture, and health care.

While the development in CPS has progressed for realizing the desired performance metrics in terms of efficiency and sustainability, it, however, has become evident that the extensive use of cyber-components may introduce vulnerabilities to these systems, allowing unwanted access to malicious intruders. As a consequence, for the safe and secure operation of CPS, consideration of cyber-security has become one of the important priorities. Security goals for CPS remain the same as those for the more conventional information technology systems. The three aspects often considered to be important for the protection of information are the following [18]: *confidentiality*, to maintain the secrecy of the important data; *integrity*, to guarantee the fidelity of the data; and *availability*, to ensure the accessibility of the data at the right time. These three security goals (sometimes abbreviated as CIA) provide a foundation for the classification of security and privacy threats in CPS.

In general, cyber-threats are categorized to the three classes of *disclosure* attacks, *deception* attacks, and *disruption* attacks [8, 106]. They can be briefly described as follows: (i) Disclosure attacks aim at acquiring sensitive and confidential information such as sensor measurements, control computation, and actuation signals. This can be achieved by passively eavesdropping on the communication in the systems. The notion of privacy is closely related to this type of attacks. (ii) Deception attacks refer to the corruption of signal data such as spoofing attacks and false data injection attacks. For CPS, this type of attacks can lead the systems to falsely operate. (iii) Disruption attacks correspond to the active intrusion of the communication by blocking or delaying the signal. Denial-of-service attacks fall in this class, which can be easily launched against both wired and wireless networks. Both the CIA goals and the security attacks have been extensively analyzed in the literature for analysis and synthesis of CPS security over the past decade [27, 106].

5.1.1 Attacks and Security in CPSs

In networked closed-loop control systems such as the one depicted in Fig. 5.1, signals are subject to attacks and threats in three main domains: during measurement, during actuation, or during the control computation. The control computation unit can be itself a network, such as a multi-agent system, where actuations, measurements, and computations are all done by the same agent, but the system is distributed and networked. Usually, in control systems, we denote the measurements and inputs at discrete time k by $y[k]$ and $u[k]$, respectively. The purpose of a control system is to take the measurements and then compute a control signal that can act on the plant through the actuators to lead to the desired performance.

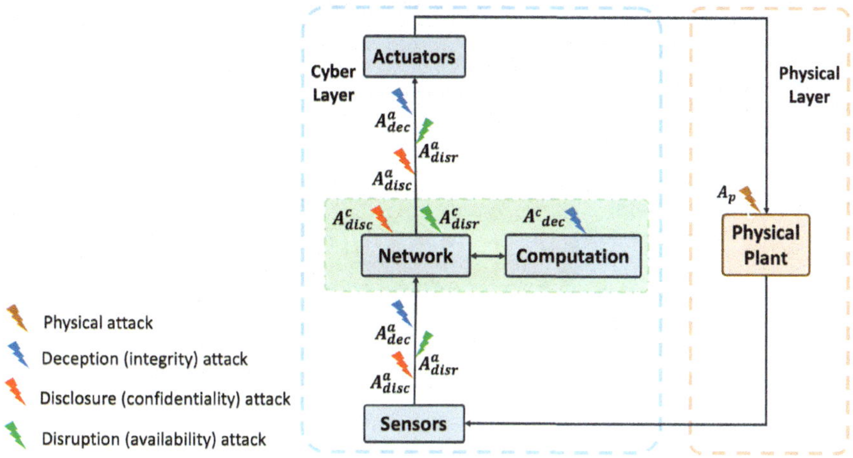

Fig. 5.1 Cyber-physical system under three types of attacks: disclosure attacks, deception attacks, and disruption attacks at different parts in the system

In a control system, where the performance goals of accuracy, speed, or robustness need to be met for valid operations, cyber-attacks can create havoc. If such a control system is part of a safety critical system, an attack could damage property and result in the loss of lives. Over the last decade, the number of cyber-attacks as well as their impact on the underlying infrastructure have steadily increased. As a result, there is a major impetus in the research community to study and understand this problem. Research efforts such as [5, 11, 93] have brought to attention the fact that the topic of cyber-attacks is not of interest only to the traditional cyber-security community and demonstrated that component-wise solutions will not suffice. Instead, these threats must be analyzed from a comprehensive system and infrastructure perspective.

5.1.2 *Faults, Attacks, and Privacy*

To obtain a systematic view of security, it is essential to model adversaries and their available resources. Figure 5.2 gives an illustration of a CPS under different types of attacks that have been discussed above. Attacks may be introduced at different parts of the system involved in communication and computation for the control. Note that physical attacks can also be made on the plant if the attacker has direct access to the system onsite. Such attacks can change the physical parameters and behaviors of the system and may even damage the system to function properly especially if actuation is applied.

Cyber-attacks can change the system from its normal behaviors, but it can often be difficult to distinguish attacks from faults in the plant or the control devices. Faults may be considered to be changes in the system that occur randomly and

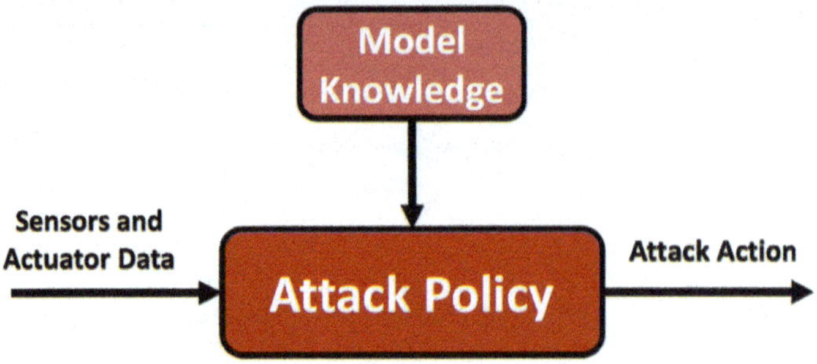

Fig. 5.2 Resources and actions of attackers targeting CPS

without particular intentions. On the other hand, attacks may be generated with specific objectives to harm the system operation. To this end, the attackers may exploit all the knowledge that they can gather regarding the system, its model, and its operation condition for finding smarter and more effective ways to attack [106]. Information useful to this effect can be obtained through publicly available sources (such as standardized control and communication protocols), eavesdropping, and even remotely breaking into the system.

Attacks against CPSs may be difficult to detect through conventional security measures based on information technology. Especially, effective attacks can be generated by only manipulating the raw data of control-related signals such as sensor measurements and command inputs. Even if the communication appears to be normal and none of the conventional safety schemes finds any problem, such manipulated data can lead to problems in the operation of the systems. For this reason, it is of great importance to develop novel methodologies and tools for CPS security based on the perspectives of systems and control.

To this end, we must first analyze the possible attack types and construct models of attacks against control systems. It is useful to categorize different attacks based on the attacker's capabilities characterized by their computational resources, the type of access they may have, the data they possess, and their capabilities to collaborate. Such a characterization is performed in the work [105], where models of attackers are provided.

In general, for considering attacks on CPS, it is critical to address issues of the following types:

- To find the potential entry points and the data of the control-related signals to which the attackers may have access.
- To identify what the attackers intend to and are capable to do to control-related signals.
- To set limitations on the capabilities of the attackers, possibly with certain safety margins.

From the defender's viewpoint, understanding these points will help in the consideration of the logic necessary for the defense mechanisms and their improvements, their level of conservatisms, and comparisons with other security techniques, which may have been already implemented.

In addition, developing attack models requires to clarify the assumptions regarding the attacks. In the area of security, the basic principle in this regard is well known as *Shannon's maxim*, which states that *the enemy knows the system* [98]. That is, security solutions should operate under the assumption that all data are shared and available to the public by direct access. *Access control* is a fundamental concept in computer security for regulating who can view and use what resources in the system [8]. In contrast, it is well known that secrecy should not be the main method to provide security in general, and thus *security by obscurity* is to be avoided.

Recently, however, privacy of signals involved in CPS is gaining much attention for avoiding the system to be overly exposed. While privacy has some roots in the area of data processing [83], in this chapter, we exclusively discuss privacy from a cyber-security perspective. It is noted that privacy and confidentiality are used interchangeably in this chapter, and attacks threatening the privacy of systems are identical to disclosure attacks.

5.1.3 Outline of This Chapter

The purpose of this chapter is to present an overview of the research activities in the area of CPS security and privacy based on the control system's approaches. We highlight recent works in this relatively new area from both the attacker's and the defender's viewpoints. For different types of attacks, we categorize the defense mechanisms based on prevention, resilience, and detection. This chapter extends the survey paper [27] on CPS security with emphasis on privacy-related issues.

The rest of this chapter is organized as follows: We first present the cyber-threat models in CPSs in Sect. 5.2 along with defense strategies effective for protection against these attacks. We then present additional techniques used to design resiliency into CPSs with techniques such as game theory in Sect. 5.3. In Sect. 5.4, we discuss the roles AI play in the area of CPS security as well as the security of AI used in CPS-related applications. We also discuss research challenges that the community will have to face in the short and mid-term for enhancing resilience against security and privacy threats.

5.2 Attack Models

In Fig. 5.1, we have presented a diagram of a typical cyber-physical system with a feedback control structure. In the physical layer is the plant to be controlled. It has the control signals $u[k]$ applied by actuators and the measured signals $y[k]$ obtained by sensors. These signals may be communicated through a network to and from the controller where sufficient computation and commands from the operators are available. We can see that there are multiple points where an adversary can make false operations on various system components and signals as indicated in the diagram. Note that there are four types of attacks shown in different colors there: physical attacks (in orange), disclosure attacks (in red), deception attacks (in blue), and disruption attacks (in green). In this section, by grouping these attacks in three categories, we discuss each of them together with examples on defense techniques for mitigating the attack impacts on the CPSs.

5.2.1 Disclosure Attacks

As the name suggests, *disclosure* attacks attempt to gain access to sensor signals and control signals in networked cyber-physical systems. Such attacks are passive in the sense that they attempt to capture readily available or leaked information by *eavesdropping* on signals exchanged between different parts of the control system on communication channels. Disclosure attacks are one of the hardest types to defend against, as in most cases they go unnoticed for long periods of time. In this sense, they can sometimes be considered as *Advanced Persistent Threats* (APTs).

After successful disclosure attacks, the attacker can use these signals to gain knowledge about the system for launching specifically engineered smart attacks on control systems. With more knowledge, the attacker may generate attacks with higher impacts to bring damages to the system. For example, the work [91] exhibits that *inference attacks* or *plaintext attacks* can be used to infer the transfer function of the system by intercepting the sensory data and control inputs; such knowledge can even be used to launch attacks specifically designed to destabilize the closed-loop system.

The defense mechanisms for disclosure attacks fall in two main categories: *cryptography* and *randomization*. Cryptography techniques encrypt the sensor and control data so that only authorized entities in the system can access them. Such techniques have been well studied in the area of computer science [57]. On the other hand, randomization techniques are grounded in control theory and are well studied for a wide range of robust control problems [73]. We discuss both of these defense techniques in the following.

5.2.1.1 Cryptography

Cryptography is the practice for secure communication by constructing and analyzing protocols that prevent third parties from reading private messages. Modern cryptography started during the World War II. The currently deployed form makes use of public and private key pairs to make sure only authorized entities have access to the encrypted data or *ciphertext*. An entity is authorized if they have access to the private key, and with it, they can decrypt the ciphertext to get the original data, i.e., the *plaintext* [28].

With public-key cryptography, each component within the CPS would need to decrypt the ciphertext before it performs the necessary computation and then encrypt it again before sending it to the next component in the control system. One critical issue in CPS is the real-time requirements in communication since even short time delays in the transmissions of control signals can easily cause performance degradation and, in the worst case, instability in feedback systems. Hence, in many control network protocols, encryption is not applied to signals having hard constraints on time delays.

Cryptography can also play an important role from the perspective of privacy. In the case of CPS, even if measurement and control signal data are encrypted during their communication over networks, they become accessible in plaintext within different components in the system. That is, the privacy of such data is not preserved. Such a system can be vulnerable to leaks as an attacker can compromise components that are vulnerable and gain access to the decrypted plaintext data there. In general, CPSs use many components, which are deployed in a range of areas. Moreover, the components may not be built to be robust for cost reasons.

Recent studies have been carried out on CPS from the viewpoints of cryptography and privacy. The work [10] studies a cryptographic framework to mitigate the threats in CPS. To address privacy issues in SCADA systems, symmetric key generation by a polynomial-based scheme is discussed in [92]. The paper [114] studies privacy assurance of the encoded data for estimating unstable systems over uncertain wiretap channels, where the zero-error secrecy capacity is found.

Homomorphic encryption is a privacy preserving form of encryption that allows computation on ciphertexts directly. By employing this technique in CPS, all measurement signals can be encrypted at the source. Then, all transmissions and subsequent control computations on the signal data can be performed on the ciphertext, and thus, its privacy is preserved. Finally, the signal can be decrypted at the plant into the plaintext for actuation.

The encryption and decryption functions for homomorphic encryption can be thought of as algebraic homomorphisms between plaintext and ciphertext spaces. Based on the classes of computation supported, there are two subclasses of homomorphic schemes. If the scheme guarantees only summation type of computations on the ciphertext, then it is called additively homomorphic. On the other hand, if it supports only multiplication type of computations, then it becomes multiplicatively homomorphic. If both of these properties hold, the encryption scheme is called fully homomorphic; otherwise, the scheme is semi-homomorphic [9, 111].

Recently, the use of homomorphic cryptography in CPS has gained attention. It was first considered in [34, 60, 63] to implement static controllers while ensuring closed-loop stability. These results were subsequently expanded to cases with dynamic controllers [17, 61, 80], estimators [4, 117], nonlinear controllers [59], and model predictive controllers [3, 21]. Other studies have considered the applications of homomorphic encryption to secure transportation systems [35] and performance optimization [67].

5.2.1.2 Randomization

An effective tool used to defend against disclosure attacks is *randomization*. With randomization techniques, we can ensure that it is nearly impossible for an adversary to reverse engineer the sensitive data from eavesdropping in the system. For instance, in [44], the idea of masking data to achieve the exact average consensus in the presence of an eavesdropper is proposed. The work [26] proposes a randomization-based method for the selection of gains to secure a control system whose sensors and actuators are under disclosure attacks. We also note that the idea of randomization has been exploited in machine learning in the presence of adversaries [49].

Here, we discuss two approaches, one based on differential privacy and the other using information-theoretic methods. A powerful framework for privacy preserving randomization is that of *differential privacy*. It can deal with problems that arise when there is sensitive data, when there is a high possibility of eavesdropping in the system, and when there are adversaries who want to recover the sensitive data. Differential privacy requires that the statistics of the publicly reported outputs of a system does not significantly change when the value of a single data is changed. This makes it hard for an adversary to reverse engineer the exact outputs or gain access to the sensitive data of the system irrespective of what additional side-channel information the adversary may have access to [32].

Differential privacy techniques have been widely applied for robustifying control systems for security. In [64], the privacy of inputs to a dynamical system originating from different participants is considered. Average consensus with differential privacy is considered in [77, 84]. Differentially private optimization has also gained considerable momentum in last few years [45, 116].

Another approach for randomization-based privacy is the information-theoretic one. Information-theoretic privacy dates back to studying secrecy and its generalizations [95]. Under this approach, as measures of privacy, entropy, mutual information, the Kullback–Leibler divergence, and the Fisher information have been repeatedly used (see [33] and the references therein). For example, in [103], the privacy of dynamical systems is considered mainly against disclosure attacks. There, privacy loss is used as a penalty component in the underlying cost function. Specifically, the penalty component is measured by the mutual information between two datasets. This approach consists of the optimization of this cost function with and without the penalty component and evaluating the resulting trade-off. We also point out that in [53], maximal leakage based on guessing theory was recently developed based on an axiomatic approach to measuring private information leakage.

5.2.2 Deception Attacks

The second class of attacks is that of deception attacks. They are integrity attacks on the sensor and control signals in CPSs. For example, False Data Injection (FDI) attempts to augment or distort the sensor and control signals so that the data values are different from their original values. Such attacks can be made at multiple locations in the closed-loop system as we have seen in Fig. 5.1. When they occur as sensor attacks, the operating conditions can be altered by biasing the sensor measurements with a non-zero value as

$$y^a[k] = y[k] + a_y[k],$$

where $y^a[k]$ is the falsified sensor measurements and $a_y[k]$ is the value augmented by the attacker on the sensor measurements [5].

Deception attacks may also target other parts of the CPS. On the actuator side, the attacks will modify the control inputs from their original signals. Similar effects can be created by computational attacks in some networked control systems; they modify the control logic so that the control input generated will become different. In either case, we can represent the attacked control input $u^a[k]$ as

$$u^a[k] = u[k] + a_u[k],$$

where $a_u[k]$ represents the value augmented by the attacker on the control inputs.

As in the case of disclosure attacks, deception attacks may be hard to detect. Especially, if the communicated information is the same except for the numerical data of sensor and control signals, then conventional security methods may not succeed in detecting such attacks. Additionally, they are active attacks, that is, they modify the underlying sensor and control signals, and hence have the potential to induce a larger impact on the control system. For example, control systems can be easily destabilized via deception attacks as demonstrated in [5, 19] for problems in the domain of power systems.

There are fundamentally three types of deception attacks as follows:

(i) injection of new signals,
(ii) interception and augmentation of signals, and
(iii) interception and replay of signals.

In the first case, the attacker injects additional signals into the control system by spoofing or posing as a sensor or control unit. For the second type, the attacker intercepts the message containing control signals and substitutes the data before placing the message back to the network. Finally, for the third type of attacks, the attacker eavesdrops on the traffic in the network and then replays them at a later time in the control system.

These attack signals add noise in the control computation, leading to incorrect stabilization points. In particular, in a replay attack, the attacker takes a recording of the signals communicated over networks under normal conditions and then uses them at a later time in their attacked signals. In this way, they may go unnoticed by detection mechanisms since the system appears to be operating as normal [78]. Replay attacks are unique in the sense that they are a mixture of disclosure, deception, and disruption types of attacks [29].

There is a range of techniques to protect control systems against deception attacks. We will discuss them in some detail next.

5.2.2.1 Outlier Detection-Based Approaches

A general principle in situations where the received data may contain errors is to collect more data and then select among those that can be trusted the most. Clearly, the number of data points to be selected should be sufficiently large for the purpose. Then, one may find outliers among the data, which differ from others and may be excluded from the subsequent processing.

As a simple example, suppose that one collects n measurements $y_i, i = 1, 2, \ldots, n$, regarding the same data, but knows that potentially up to f of them can be corrupted. If the bound f is smaller than $n/2$, one may use majority voting to identify faulty values or outliers, so as to find the true data or its estimate. In statistics, outlier detection is a basic problem, and various robust methods have been formulated as, e.g., robust versions of least squares problems [71].

Such a principle can be found in control problems. The paper [36] considers an estimation problem for a linear system under adversarial sensor attacks. The system is equipped with n sensors but up to f of them may not transmit their true measurements. A fundamental limitation is identified, showing that if more than half of the sensors are attacked, safe estimation of the state is impossible. Then, a specific state reconstructor is proposed, which is resilient to attacks. There, a bank of observers is used by the system based on ideas from compressed sensing and error corrections over the reals. Moreover, for a control problem using output feedback, it is shown that the separation principle of estimation and control holds even if the sensors are attacked. This work has inspired many related research in the area including [74, 99].

5.2.2.2 MSR Approach in Multi-agent Consensus

Another related approach has been studied in the context of multi-agent consensus. There, n agents in a network collaborate to arrive at a consensus in their state values by exchanging them in an interactive manner. In resilient consensus problems, up to f agents may be faulty or attacked and transmit arbitrary state values to their neighbors. Again, as in the outlier detection approach, the knowledge of f is assumed available to the defenders. Resilience in this class of problems can be attained by the so-called *Mean Subsequence Reduced (MSR)* algorithms. The underlying idea is relatively simple since the agents ignore suspicious values among those received from other agents and compute their new states at each time they make updates. This class of algorithms is effective against adversarial agents of the Byzantine types (e.g., [68]) and has recently been studied from both computer science and control perspectives [25, 65, 66, 68, 118].

More specifically, under the assumption that a bound f on the number of malicious agents in the entire system is known, each agent treats as outliers the f largest values as well as the f smallest values among those received from neighboring agents. As mentioned above, the approach is simple, and as a consequence, the agents are not capable to fully identify the adversaries. Nevertheless, the nonfaulty agents can reach a consensus at a value without deviating too much from their original initial states.

It is notable that the necessary level of connectivity in the network structure can be fully characterized for this problem. In particular, the relevant topological notion in this problem is called *graph robustness*, introduced by [66]. It represents how well groups of agents are connected over different paths. As the definition involves combinatorial issues, it is computationally hard to determine the robustness of a given graph [118]. However, to obtain bounds on robustness, mixed integer linear programming can be utilized in an efficient way [108].

MSR algorithms have close relations with problems in distributed algorithms in computer science. Studies from control theoretic viewpoints include those on consensus [24, 25, 66], distributed state estimation [75], clock synchronization [58], and distributed optimization [101]. Another application of the Byzantine consensus is recently receiving much attention in computer science, which is related to blockchain and cryptocurrencies (see, e.g., [42]).

5.2.2.3 Trust-Based Approaches

Trust-based approaches have been developed to enhance resilience of multi-agent systems. In systems involving multiple agents capable to make their own decisions based on local information, agents may determine the degree of reliability about their neighbor agents depending on how they interact over time. If some of the neighbors are found not trustworthy, agents may try to reduce the influence that they receive from such agents.

For example, in [46], consensus problems are considered, where the number of untrustworthy agents is assumed to be small so that sufficiently many paths remain in the network connecting trusted agents for forming the consensus among themselves. There, a function of trust is introduced, which is updated by the agents as they interact with each other. A similar algorithm for computing trust indices is used for a distributed estimation problem in [55].

Such a trust-based approach can be useful in more adversarial environments as well. The work [1] improves the performance of MSR-based consensus by introducing trusted nodes in the network. Such nodes are built more resilient to attacks and hence are to be considered always nonfaulty and reliable. Trust-based approaches have mainly been developed for security against deception attacks and especially for the protection of sensor networks [2]. A detailed survey can be found in [79] regarding trust models in sensor networks in terms of data trust and communication trust.

5.2.2.4 Other Approaches

The three approaches against deception attacks discussed above are a small set of samples in view of the variety of studies in this new but active research domain. In what follows, to highlight this point, we mention a few more studies.

The paper [97] studies a security problem of a linear dynamical system and discusses the significance of the notions of securable and unsecurable subspaces of the state space. This is analogous to the controllable and unobservable subspaces of the system. Certain guarantees can be maintained with respect to the performance of estimating the states in the securable subspace.

We would also like to mention the work [26]. It proposes an information retrieval approach for the defense when the cyber-layer is under deception attacks. At each time step, the controller is able to find and make use of healthy and unattacked data, which is then used in the state feedback. A distributed method for the information retrieval is investigated.

Security issues have been extensively looked at from the signal processing viewpoint as well. In [37], it is discussed that security issues can arise in the use of GPS when intentional interferences are created by jamming. Such situations can occur when hijackers on a stolen vehicle try to avoid being tracked. One promising solution is to employ receiver architectures based on multiple antennas with robust signal processing techniques.

5.2.3 Disruption Attacks

Disruption attacks refer to attacks that interrupt the communication of information intentionally. This class of attacks can be realized through *Denial-of-Service* (DoS) attacks and *jamming* attacks. Such attacks can be launched to target measurement data and actuation data communicated over different networks [12, 52]. It is not particularly difficult to generate such attacks once the adversaries have access to the channel. For DoS attacks, many packets must be sent to the same destination with short intervals while for jamming attacks, one must generate sufficient interference signals to disable wireless communication.

When the network is under disruption attacks, the transmitted data does not arrive at the receiver's end. To compensate the missing data, one approach is to replace it with the value of zero. Thus, the attacked signals $y^a[k]$ and $u^a[k]$ may be expressed as the following:

$$y^a[k] = \begin{cases} 0 & \text{if the measurement } y[k] \text{ is attacked;} \\ y[k] & \text{otherwise.} \end{cases}$$

$$u^a[k] = \begin{cases} 0 & \text{if the control input } u[k] \text{ is attacked;} \\ u[k] & \text{otherwise.} \end{cases}$$

Another approach is to keep the measurement $y[k-1]$ or the control input $u[k-1]$ from the previous time step; this approach may be effective if the attacks are short in time.

It is obvious that if the attacks continue for a long period of time, the control of the system will be lost, which will result in performance degradation and even instability of the overall system. To consider more concrete situations, there have been attempts to introduce some constraints in attack models. In particular, the works [13, 23, 100] employed models of DoS attacks where the time duration and/or the frequencies of the attacks are limited. The motivation comes from the constraints of energy resources that the attacker may have for launching such attacks. The frequency and time duration of attacks clearly affect the energy consumption especially if the attacker uses battery-driven devices to emit interference signals against wireless communication.

Note that DoS attacks can be viewed as data losses during the communication. Such losses can also occur in normal conditions from data packets being overly delayed or corrupted due to noise in the channels. In communication theory, such interferences have been modeled using probabilistic models. There, data losses occur randomly, which can be represented by simple models such as the Bernoulli models and the Markov chain models.

On the other hand, for DoS attacks, such probabilistic models will be insufficient. This is because the adversaries may have more information about the targeted system and may generate attacks to maximize the impacts even if he is limited in the energy budget. Hence, it is of importance to assume the worst-case model for the attack patterns and the model necessarily becomes deterministic.

In the above-mentioned works, conditions on the attack model parameters are derived for the closed-loop system to achieve stability. These results are extended to the case where the channels are not only subject to DoS attacks but also to random data losses in [13, 14]. In this case, the analysis approach becomes more complicated as a stochastic framework is necessary where the attacks are represented by a process whose distribution is partially unknown.

Several measures to mitigate the damages potentially caused by DoS-type attacks have been developed. As the transmissions may be unsuccessful in face of the attacks, it helps the controller to adjust the transmission times according to the level of attacks. One such mechanism can be realized by *event-triggered* control schemes (see, e.g., [48]). Under such schemes, data transmission times are determined in real time by comparing the signal values to be sent with those successfully received the last time. When the difference is larger than a given threshold and attacks are present, the transmission frequency may be increased [23, 31, 100]. Alternatively, the transmission times can be randomly chosen as well, which is the approach proposed in [13]. As we have seen earlier, randomization can be a useful method in security. In this case, it makes it difficult for the attacker to predict the transmission times.

5.3 Common Techniques Against Deception and Disruption

Among the defense mechanisms for CPS security, some are not against only specific types of attacks. We divide these mechanisms into two main categories: resilience and detection.

5.3.1 Resilience Mechanisms

The original definition of the word *resilience* is the ability to rapidly return to good condition after troubles and disruptions occur. In the context of cyber-security, its meaning is somewhat extended and is the ability to prepare for and adapt to various conditions and to withstand and recover from severe stresses and deliberate attacks [7, 36]. Control systems may not inherently possess resilience against cyber-attacks, and hence, special attention must be paid to enhance their resilience properties. In this subsection, we make a brief introduction to game theory, which provides important tools and viewpoints for developing resilience mechanisms against deception and disruption attacks.

5.3.1.1 Game-Theoretic Methods

To consider resilience problems based on game-theoretic approaches, we must clarify the gain and cost of the adversary for attacking the system as well as the damage and the cost of the defender, which may correspond to the system operator. It is clear that there are trade-off relations for both players at various levels, also depending on each other's strategies. Game theory provides a general framework to study the interaction between the two players of the adversary and the defender. Each player has a utility function, which is to be optimized. However, the utility depends also on the actions of the other player, and hence, each player must take account of the strategies of the opponent. This is the challenging aspect of problems in games.

A number of game-theoretic problems have been studied for enhancing security levels of CPSs. From the viewpoint of game theory, there are mainly two important aspects specific to security problems. One is due to the structure of the CPS and the other depends on the types of actions in the cyber-layer. We briefly explain these two aspects as follows.

The first issue specific to CPS security problems comes from the structure of CPS consisting of the cyber-layer and the physical layer [6, 15, 20, 94, 120]. There, games can be formulated in both physical- and cyber-layers, and as a result, the concept of *games-in-games* can be introduced to highlight the cross-layer structure involving two interconnected games, one in the physical layer and the other in the cyber-layer.

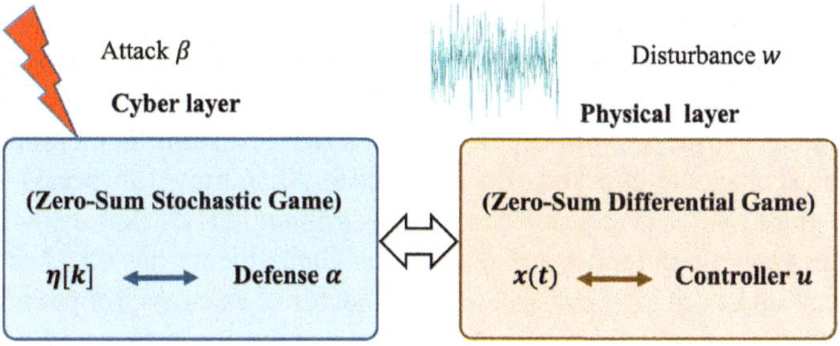

Fig. 5.3 Games-in-games concept involving both physical and cyber-layers

For the physical layer, the system dynamics can be expressed by the general nonlinear model in continuous time

$$\dot{x}(t) = g(t, x, u, w, \eta(t, \alpha, \beta)), \tag{5.1}$$

where $g(\cdot)$ is a nonlinear function of the state x, the control input u, and the disturbance w, and t is the time index. In this model, the decisions made in the cyber-layer are represented by $\eta(t, \alpha, \beta)$, where α and β are the actions of the attacker and the defender, respectively. The dynamics in the cyber-layer may be discrete in nature and the actions α and β may be represented by the Markov chains in discrete time with state transition models. As a consequence, in Fig. 5.3, the overall system can be seen as a hybrid system.

For instance, at the physical layer, we can formulate a robust control differential game between the controller and the disturbance. This will lead us to an $\mathcal{H}_\infty$ design of the controller for achieving robust performance for disturbances and uncertain parameters in the system [86]. The cyber-layer consists of the defense system, and thus, a game between the defender and the attacker can be formulated to obtain optimal cyber-policies to guarantee the security of the overall control system [119].

The second aspect to be considered in CPS security problems is based on the types of attacks and malicious behaviors. More specifically, we can classify and formulate different classes of games depending on whether the attackers and the defenders are active or passive. For example, the defender can be passive in responding to the attacks. Then, the defender's interaction with the attacker can be reasonably captured by a Stackelberg game; there, the attacker is active in sending interference signals at a targeted level to reduce the communication quality while after observing the attack strategy, the defender rationally determines whether/how to increase the transmission power to overcome and mitigate the influence of the interference. On the other hand, the defender can be prepared to defend the system in a more active manner if, for example, he is more informed of the attacker's strategy. In such cases, the Nash equilibria form a more reasonable class of solution concept [43]. For example, the work [109] considers a two-player game with an attacker and a defender that play over a network whose topology is unknown. The defender attempts to achieve synchronization among the network nodes by injecting control inputs while reducing the influence of the worst-case attack signal generated by adversarial nodes.

We discuss a few notable recent results on CPS security based on game-theoretic approaches. In [110], a zero-sum game is considered for an estimation problem where some of the sensors are attacked. The works [16, 90] address the threats for control systems using cloud services and formulate signaling games that explicitly model the trust between the defender and the threats. Game theory has also been used for studying eavesdropping in [70], where the behavior of an eavesdropper is viewed as that of a follower in a Stackelberg game against a user who employs active defenses.

5.3.2 *Detection Mechanisms*

A detection mechanism usually uncovers the existence of an attack by monitoring its effects on the outputs of the system. In addition to detecting the existence of an attack, stronger strategies can be installed to *identify* (or *localize*) the set of nodes/signals that are attacked (e.g., [88]). If the effect of the attack signals cannot be traced by the outputs, these attacks are *covert* [106] or *stealthy* [104]. The survey papers [29, 40] have reviewed some detection mechanisms for deception, as well as disruption attacks in CPS.

In what follows, we classify all detection and recovery methods proposed in the controls literature into four categories which include observer-based techniques, watermarking, baiting, and learning-based anomaly detection.

5.3.2.1 Observer-Based Techniques

It is common to design observers for systems having unmeasured states so as to estimate them in a systematic manner. For the detection of cyber-attacks, observers can play a pivotal role. One can compute the expected output of the system based on the control input applied at the time. Such an estimate can be compared with the sensor measurement, and then we can focus on any difference between them, which is called the *residues*. A detection alarm may be set off in the event where the residues become greater than a given threshold [105]. Such detectors have been traditionally studied in systems control under the subject of fault detection. Common methods in this area include those based on geometric control and those using the so-called unknown input observers [22, 72]. In certain cases, these methods enable the system operators to identify where in the system, unknown attack signals could have been injected.

This approach has been generalized to the CPS setting in [88]. It is pointed out that large-scale CPS models such as power systems involve linear algebraic conditions. The above-mentioned reference considers deception and disruption attacks on both sensors and actuators and develops a systematic method to design filters for detection and identification of attacks in both centralized and distributed settings. Though the detection of attacks can be relatively easy, it is found that the identification of the attack location is computationally hard. The work [89] addresses this issue and proposes algorithms for efficiently carrying out detection and identification by distributed and decentralized monitors. In particular, the method is based on a divide-and-conquer principle for reducing the computational burden.

As a related work, we mention [81], which studies the impact of sensor attacks on networked control systems by measuring the reachable set of the attacks. Similar ideas have been used in the context of multi-agent systems in the presence of misbehaving nodes [87, 102], where each agent is equipped with a filter for fault detection and identification of the network.

While these observer-based approaches provide useful, secure methods for CPS, they require high computational resources since various matrices must be used for estimating outputs and detecting attacks. In [99], an efficient version of these methods is built by using a multi-modal observer to decrease the complexity of the filters. The attack-free sensors can be identified for generating correct state estimates, which employ Satisfiability Modulo Theory to support the reasoning.

5.3.2.2 Watermarking

A digital watermark is a hidden information that is embedded in digital signals to show copyright ownership or authentication of such signals. As in the traditional watermark in paper, the objective is to discourage counterfeiting. This technique has been used for the purpose of CPS security to detect attacks [76]. The approach is to add watermark signals in the control-related signals transmitted over networks. The watermarks are perceptible at the receiving ends, and thus the controllers and actuators can verify that the signals are authentic and not falsified by attackers. In particular, watermarking is known to be effective in detecting replay attacks; since such attacks simply replace the current control signals with recorded ones from the past, they can be detected by embedding time information.

Another control-oriented usage of watermarking for detecting sensor deception attacks has been reported by [96]. In this approach named *dynamic watermarking*, each actuator injects its own watermark signal into its control input. Since the control input is applied to the plant, the watermarks in fact become physical ones. The sensors attempt to find the effects of the physical watermarks in the measured signals and report those to the controller. If any of the reports from the sensors contain errors, it might be an indication of the sensors being compromised. It is shown using linear systems under the Gaussian noises that for the attacker to remain undetected by this method, the potential damages caused by sensor attacks will be constrained. The effectiveness of dynamic watermarking is demonstrated in the application domains of power systems in [51] and transportation systems in [62].

In addition, the work [112] considered a CPS problem where watermarks are generated by both adding the Gaussian signals and randomly dropping packets based on the Bernoulli processes. For detecting deception attacks, a correlation detector is designed, and the trade-offs between the security level and control performance degradation is studied.

5.3.2.3 Baiting and Moving Targets

An interesting CPS security question can be formulated as follows: In a worst-case situation potentially facing a powerful attacker who has full knowledge of the entire networked control system as well as full access to the sensors and actuators, how can the system operator actively secure the system and detect the presence of such an attacker? That is, how can one lead the attacker so that he cannot remain to be stealthy? Such questions have been studied in the literature where resilient methods against such worst-case scenarios are designed.

The method proposed in [38] is based on baiting by using a cyber-security tool called *honeypot* or more specifically a *honeytoken*. This is a piece of data made accessible to the attacker, and by the attacker probing the honeytoken, he might unintentionally inform the defender of the attack. In particular, the approach is to make a false model of the system dynamics available to the attacker. Then, even if stealthy attacks are generated based on the false but carefully designed model, the system can not only detect the attacks but also operate safely based on the attacked signals.

An alternative approach to deceive the attacker is proposed in [113] based on the idea of *moving target*. Here, virtual dynamics is introduced in the system, which is time varying depending on the original system states and, more important, securely hidden from the attacker. The presence of this virtual system disables the attacker to correctly identify the system even if he has full access to the measurements and control inputs. Moreover, even if the attacker generates stealthy attacks based on the false model, the defender will be able to detect the attacks. It is critical that the virtual system is time varying to make it difficult for the attacker to learn the system accurately over time. This approach has been extended in [113] to detect attacks on both the control inputs and sensor measurements.

In the general area of cyber-security [54], moving target defenses refer to security schemes where system attributes are changed by the defender with the intention to make the attack surface unpredictable.

5.3.2.4 Learning-Based Anomaly Detection

Here, we very briefly discuss the machine learning-based technique of *anomaly detection*. It is for the detection of outlying data whose properties and patterns do not match with other data in the dataset [82]. Anomaly detection is widely used for intrusion detection in the area of computer networks (see, e.g., the review articles [39, 107]). Learning based on neural networks has also been employed in anomaly detection in the context of cyber-security of power systems and their control systems [47].

5.3.2.5 Other Approaches

There are several studies that discuss designing anomaly detectors through other tools. A recursive distributed Kalman filter in the presence of sensor attacks is developed in [30, 74]. Some studies combine the resiliency with detection methods and investigate the resilience of the detection tools under attacked conditions (e.g., [85]).

5.4 Summary and Future Directions

High integration of instrumentation, communication, and control into physical systems is currently enhancing the operability and efficiency of large-scale systems in various areas. This trend in the industry has led the study of CPS with increased attention. As we have discussed in this chapter, a critical feature ubiquitous in CPS is the need to ensure their security in the face of adversarial cyber-attacks. Unlike faults, cyber-attacks correspond to system-specific, Malicious, and active threats launched by attackers with the knowledge of the system and strong motivations to maximize the impact on the CPS. This chapter has been arranged in a tutorial fashion to serve as an introduction to this area. Next, we discuss several directions which may be important for future studies.

5.4.1 CPS-Security of AI-Driven CPSs

The area of Artificial Intelligence (AI) has experienced twists and turns as it has seen several ethical and technical challenges over the decades. Many techniques have emerged and proved useful in some applications, but also showed their weaknesses and flaws. With significant improvements in computational power and increase in the volume of available data, AI has recently made significant advances. It is expected that AI will serve as the core technology in safety critical domains such as transportation systems.

One technical reason for the difficulty is that AI also suffers from cyber-vulnerabilities. The current AI techniques such as deep learning are highly vulnerable to input modifications. For example, in image recognition applications based on Neural Networks (NNs), by modifying an image of a cat with subtle perturbation which may be imperceptible to human eyes, it can be identified as a dog. The work [56] describes how a non-resilient classifier NN can be against deception attacks in both training and inference stages. Likewise, other studies demonstrate the weaknesses of reinforcement learning tools against attacks (see, e.g., [50]).

Cyber-physical systems in their perception or autonomy may use these AI techniques, but until these issues are not resolved, it might be difficult to introduce a broad AI utilization in the industry. The work [115] reviews comprehensively the vulnerabilities in AI and defense approaches against potential attacks. Adversarial machine learning [49] and robust learning [69] are two examples of resilient mechanisms for the protection of AI and reduction of their vulnerabilities. There are a lot of open and essential problems in this domain which have to be solved by new generation AI.

5.4.2 AI for Cyber-Security

While AI can be a subject to cyber-attacks, it has been widely used for the detection of various anomalies and faults in data. Along this direction, AI has been extended to detect adversarial targeting CPS data. A deeper and stronger usage of AI, rather than a binary anomaly detector, is an interesting domain of research. For example, techniques that require less data, such as the statistical Bayesian inference and representation learning [41], may be better options in case the number of samples due to attacks may be small.

5.4.3 Scalability

Like other hard problems arising in complex systems, scalability is a major concern in cyber-security problems and solutions. Many trivial solutions such as hardening or backup/redundant systems suffer from a lack of scalability. Scalability should be taken into account in all cyber-security problem formulations. Researchers with the theoretical background may overlook this aspect.

5.4.4 Toward More Realistic Cyber-Security Solutions

The development of CPS security solutions is often about addressing intentional faults targeting cyber-components in CPS to degrade and damage the functionalities of the systems. Random faults are different in that they are not intentional and happen without a designer (though they may result from human errors). In all cyber-security risk assessments, smart but yet feasible intentions, knowledge, and access of attackers must be taken into full account.

At the same time, defense mechanisms should be designed with realistic understanding of the impact of cyber-attacks. In fact, safety mechanisms of CPS must address a range of problems at different levels in the systems. On the other hand, all cyber-security problems are safety problems, but not vice versa. Thus, we need a significantly stronger case in dealing with CPS security. In particular, the authors see room for improvement in cyber-security threat assessment, which should differ from safety risk assessment of systems.

Acknowledgements We would like to thank Farhad Farokhi for several useful discussions on the privacy of cyber-physical systems.

References

1. Abbas, W., Laszka, A., Koutsoukos, X.: Improving network connectivity and robustness using trusted nodes with application to resilient consensus. IEEE Trans. Control Netw. Syst. **5**(4), 2036–2048 (2017)
2. Ahmed, A., Bakar, K.A., Channa, M.I., Haseeb, K., Khan, A.W.: A survey on trust based detection and isolation of malicious nodes in ad-hoc and sensor networks. Front. Comp. Sci. **9**(2), 280–296 (2015)
3. Alexandru, A.B., Morari, M., Pappas, G.J.: Cloud-based MPC with encrypted data. In: Proceedings of IEEE Conference on Decision and Control, pp. 5014–5019 (2018)
4. Alexandru, A.B., Pappas, G.J.: Encrypted LQG using labeled homomorphic encryption. In: Proceedings of ACM/IEEE International Conference on Cyber-Physical Systems, pp. 129–140 (2019)
5. Amin, S., Schwartz, G.A., Hussain, A.: In quest of benchmarking security risks to cyber-physical systems. IEEE Netw. **27**(1), 19–24 (2013)
6. Amin, S., Schwartz, G.A., Sastry, S.S.: Security of interdependent and identical networked control systems. Automatica **49**(1), 186–192 (2013)
7. Annaswamy, A.M., Malekpour, A.R., Baros, S.: Emerging research topics in control for smart infrastructures. Annu. Rev. Control. **42**, 259–270 (2016)
8. Bishop, M.: Introduction to Computer Security. Addison-Wesley (2005)
9. Brakerski, Z., Gentry, C., Vaikuntanathan, V.: Fully homomorphic encryption without bootstrapping. ACM Trans. Comput. Theory **6**(3), 1–36 (2014)
10. Burmester, M., Magkos, E., Chrissikopoulos, V.: Modeling security in cyber-physical systems. Int. J. Crit. Infrastruct. Protect. **5**(3–4), 118–126 (2012)
11. Cárdenas, A.A., Amin, S., Lin, Z.S., Huang, Y.L., Huang, C.Y., Sastry, S.: Attacks against process control systems: Risk assessment, detection, and response. In: Proceedings of ACM Symposium Information, Computer and Communications Security, pp. 355–366 (2011)
12. Cárdenas, A.A., Amin, S., Sastry, S.: Secure control: Towards survivable cyber-physical systems. In: Proceedings of 28th International Conference on Distributed Computing Systems Workshops, pp. 495–500 (2008)
13. Cetinkaya, A., Ishii, H., Hayakawa, T.: Networked control under random and malicious packet losses. IEEE Trans. Autom. Control **62**(5), 2434–2449 (2017)
14. Cetinkaya, A., Ishii, H., Hayakawa, T.: An overview on Denial-of-Service attacks in control systems: Attack models and security analyses. Entropy **21**(2), 210 (2019)
15. Chen, J., Zhu, Q.: Resilient and decentralized control of multi-level cooperative mobile networks to maintain connectivity under adversarial environment. In: Proceedings of IEEE Conference on Decision and Control, pp. 5183–5188 (2016)
16. Chen, J., Zhu, Q.: Security as a service for cloud-enabled internet of controlled things under advanced persistent threats: a contract design approach. IEEE Trans. Inf. Forensics Sec. **12**(11), 2736–2750 (2017)
17. Cheon, J.H., Han, K., Kim, H., Kim, J., Shim, H.: Need for controllers having integer coefficients in homomorphically encrypted dynamic system. In: Proceedings of IEEE Conference on Decision and Control, pp. 5020–5025 (2018)
18. Cherdantseva, Y., Hilton, J.: A reference model of information assurance & security. In: Proceedings of International Conference on Availability, Reliability and Security, pp. 546–555 (2013)
19. Chockalingam, N., Chakrabortty, A., Hussain, A.: Mitigating denial-of-service attacks in wide-area LQR control. In: Proceedings of IEEE Power and Energy Society General Meeting, pp. 1–5 (2016)

20. Clark, A., Zhu, Q., Poovendran, R., Baar, T.: An impact-aware defense against Stuxnet. In: Proceedings of American Control Conference, pp. 4140–4147 (2013)
21. Darup, M.S., Redder, A., Shames, I., Farokhi, F., Quevedo, D.: Towards encrypted MPC for linear constrained systems. IEEE Control Systems Letters **2**(2), 195–200 (2018)
22. De Persis, C., Isidori, A.: A geometric approach to nonlinear fault detection and isolation. IEEE Trans. Automatic Control **46**(6), 853–865 (2001)
23. De Persis, C., Tesi, P.: Input-to-state stabilizing control under Denial-of-Service. IEEE Trans. Automatic Control **60**(11), 2930–2944 (2015)
24. Dibaji, S.M., Ishii, H.: Resilient consensus of second-order agent networks: Asynchronous update rules with delays. Automatica **81**, 123–132 (2017)
25. Dibaji, S.M., Ishii, H., Tempo, R.: Resilient randomized quantized consensus. IEEE Trans. Automatic Control **63**(8), 2508–2522 (2018)
26. Dibaji, S.M., Pirani, M., Annaswamy, A.M., Johansson, K.H., Chakrabortty, A.: Secure control of wide-area power systems: Confidentiality and integrity threats. In: Proc. IEEE Conf. Decision and Control, pp. 7269–7274 (2018)
27. Dibaji, S.M., Pirani, M., Flamholz, D.B., Annaswamy, A.M., Johansson, K.H., Chakrabortty, A.: A systems and control perspective of CPS security. Annu. Rev. Control. **47**, 394–411 (2019)
28. Diffie, W., Hellman, M.: New directions in cryptography. IEEE Trans. Information Theory **22**(6), 644–654 (1976)
29. Ding, D., Han, Q.L., Xiang, Y., Ge, X., Zhang, X.M.: A survey on security control and attack detection for industrial cyber-physical systems. Neurocomputing **275**, 1674–1683 (2018)
30. Ding, K., Li, Y., Quevedo, D.E., Dey, S., Shi, L.: A multi-channel transmission schedule for remote state estimation under DoS attacks. Automatica **78**, 194–201 (2017)
31. Dolk, V.S., Tesi, P., De Persis, C., Heemels, W.P.M.H.: Event-triggered control systems under Denial-of-Service attacks. IEEE Trans. Control of Network Systems **4**(1) (2017)
32. Dwork, C.: Differential privacy. In: Encyclopedia of Cryptography and Security, pp. 338–340. Springer (2011)
33. Farokhi, F., Sandberg, H.: Fisher information as a measure of privacy: Preserving privacy of households with smart meters using batteries. IEEE Trans. Smart Grid **9**(5), 4726–4734 (2018)
34. Farokhi, F., Shames, I., Batterham, N.: Secure and private control using semi-homomorphic encryption. Control. Eng. Pract. **67**, 13–20 (2017)
35. Farokhi, F., Shames, I., Johansson, K.H.: Private and secure coordination of match-making for heavy-duty vehicle platooning. In: Proc. IFAC World Congress, pp. 7345–7350 (2017)
36. Fawzi, H., Tabuada, P., Diggavi, S.: Secure estimation and control for cyber-physical systems under adversarial attacks. IEEE Trans. Automatic Control **59**(6), 1454–1467 (2014)
37. Fernandez-Prades, C., Arribas, J., Closas, P.: Robust GNSS receivers by array signal processing: Theory and implementation. Proc. IEEE **104**(6), 1207–1220 (2016)
38. Flamholz, D.B., Annaswamy, A.M., Lavretsky, E.: Baiting for defense against stealthy attacks on cyber-physical systems. In: Proc. AIAA Scitech Forum, pp. 23–38 (2019)
39. Garcia-Teodoro, P., Diaz-Verdejo, J., Maciá-Fernández, G., Vázquez, E.: Anomaly-based network intrusion detection: Techniques, systems and challenges. Computers & Security **28**(1–2), 18–28 (2009)
40. Giraldo, J., Urbina, D., Cardenas, A., Valente, J., Faisal, M., Ruths, J., Tippenhauer, N.O., Sandberg, H., Candell, R.: A survey of physics-based attack detection in cyber-physical systems. ACM Computing Surveys **51**(4) (2018)
41. Goodfellow, I., Bengio, Y., Courville, A.: Deep Learning. MIT press (2016)
42. Gramoli, V.: From blockchain consensus back to Byzantine consensus. Futur. Gener. Comput. Syst. **107**, 760–769 (2020)
43. Gupta, A., Langbort, C., Baar, T.: Optimal control in the presence of an intelligent jammer with limited actions. In: Proc. IEEE Conf. Decision and Control, pp. 1096–1101 (2010)
44. Gupta, N., Katz, J., Chopra, N.: Privacy in distributed average consensus. In: Proc. IFAC World Congress, pp. 9515–9520 (2017)

45. Han, S., Topcu, U., Pappas, G.J.: Differentially private distributed constrained optimization. IEEE Trans. Automatic Control **62**(1), 50–64 (2016)
46. Haus, T., Palunko, I., Tolić, D., Bogdan, S., Lewis, F.L., Mikulski, D.G.: Trust-based self-organising network control. IET Control Theory & Applications **8**(18), 2126–2135 (2014)
47. He, Y., Mendis, G.J., Wei, J.: Real-time detection of false data injection attacks in smart grids: A deep learning-based intelligent mechanism. IEEE Trans. Smart Grid **8**(5), 1–12 (2016)
48. Heemels, W., Johansson, K.H., Tabuada, P.: An introduction to event-triggered and self-triggered control. In: Proc. IEEE Conf. Decision and Control, pp. 3270–3285 (2012)
49. Huang, L., Joseph, A.D., Nelson, B., Rubinstein, B.I.P., Tygar, J.D.: Adversarial machine learning. In: Proc. ACM Workshop on Security and Artificial Intelligence, pp. 43–58 (2011)
50. Huang, S., Papernot, N., Goodfellow, I., Duan, Y., Abbeel, P.: Adversarial attacks on neural network policies. arXiv preprint arXiv:1702.02284 (2017)
51. Huang, T., Satchidanandan, B., Kumar, P., Xie, L.: An online detection framework for cyber attacks on automatic generation control. IEEE Trans. Power Systems **33**(6), 6816–6827 (2018)
52. Hussain, A., Amin, S.: NCS security experimentation using DETER. In: Proc. 1st Int. Conf. High Confidence Networked Systems, pp. 73–80 (2012)
53. Issa, I., Wagner, A.B., Kamath, S.: An operational approach to information leakage. IEEE Trans. Inf. Theory **66**(3), 1625–1657 (2020)
54. Jajodia, S., Ghosh, A.K., Swarup, V., Wang, C., Wang, X.S.: Moving Target Defense: Creating Asymmetric Uncertainty for Cyber Threats. Springer (2011)
55. Jiang, T., Baras, J.S.: Trust evaluation in anarchy: A case study on autonomous networks. In: Proc. IEEE Int. Conf. Computer Communications, pp. 1–12 (2006)
56. Johnson, A., Grumbling, E.: Proc. Workshop on Implications of Artificial Intelligence for Cybersecurity. National Academies of Sciences, Engineering, and Medicine (2019)
57. Katz, J., Menezes, A.J., Van Oorschot, P.C., Vanstone, S.A.: Handbook of Applied Cryptography. CRC Press (1996)
58. Kikuya, Y., Dibaji, S.M., Ishii, H.: Fault-tolerant clock synchronization over unreliable channels in wireless sensor networks. IEEE Trans. Control Netw. Syst. **5**(4), 1551–1562 (2018)
59. Kim, J., Farokhi, F., Shames, I., Shim, H.: Toward nonlinear dynamic control over encrypted data for infinite time horizon. In: Proc. IFAC World Congress (2020)
60. Kim, J., Lee, C., Shim, H., Cheon, J.H., Kim, A., Kim, M., Song, Y.: Encrypting controller using fully homomorphic encryption for security of cyber-physical systems. IFAC-PapersOnLine **49**(22), 175–180 (2016)
61. Kim, J., Shim, H., Han, K.: Comprehensive introduction to fully homomorphic encryption for dynamic feedback controller via LWE-based cryptosystem. In: Privacy in Dynamical Systems, pp. 209–230. Springer (2020)
62. Ko, W.H., Satchidanandan, B., Kumar, P.: Theory and implementation of dynamic watermarking for cybersecurity of advanced transportation systems. In: Proc. IEEE Conf. Communications and Network Security, pp. 416–420 (2016)
63. Kogiso, K., Fujita, T.: Cyber-security enhancement of networked control systems using homomorphic encryption. In: Proc. IEEE Conf. Decision and Control, pp. 6836–6843 (2015)
64. Le Ny, J., Pappas, G.J.: Differentially private filtering. IEEE Trans. Automatic Control **59**(2), 341–354 (2013)
65. LeBlanc, H.J., Koutsoukos, X.: Resilient first-order consensus and weakly stable, higher order synchronization of continuous-time networked multiagent systems. IEEE Trans. Control of Network Systems **5**(3), 1219–1231 (2018)
66. LeBlanc, H.J., Zhang, H., Koutsoukos, X., Sundaram, S.: Resilient asymptotic consensus in robust networks. IEEE J. Sel. Areas Commun. **31**, 766–781 (2013)
67. Lu, Y., Zhu, M.: Privacy preserving distributed optimization using homomorphic encryption. Automatica **96**, 314–325 (2018)
68. Lynch, N.A.: Distributed Algorithms. Morgan Kaufmann (1996)
69. Madry, A., Makelov, A., Schmidt, L., Tsipras, D., Vladu, A.: Towards deep learning models resistant to adversarial attacks. arXiv preprint arXiv:1706.06083 (2017)

70. Manshaei, M., Zhu, Q., Alpcan, T., Basar, T., Hubaux, J.P.: Game theory meets network security and privacy. ACM Comput. Surv. **45**, 53–73 (2013)
71. Maronna, R.A., Martin, R.D., Yohai, V.J.: Robust Statistics: Theory and Methods. Wiley (2006)
72. Massoumnia, M.A., Verghese, G.C., Willsky, A.S.: Failure detection and identification. IEEE Trans. Automatic Control **34**(3), 316–321 (1989)
73. Milanese, M.: Robustness in Identification and Control. Springer (2013)
74. Mishra, S., Shoukry, Y., Karamchandani, N., Diggavi, S.N., Tabuada, P.: Secure state estimation against sensor attacks in the presence of noise. IEEE Trans. Control Netw. Syst. **4**(1), 49–59 (2017)
75. Mitra, A., Sundaram, S.: Distributed observers for LTI systems. IEEE Trans. Automatic Control **63**(11), 3689–3704 (2018)
76. Mo, Y., Hespanha, J.P., Sinopoli, B.: Resilient detection in the presence of integrity attacks. IEEE Trans. Signal Processing **62**(1), 31–43 (2014)
77. Mo, Y., Murray, R.M.: Privacy preserving average consensus. IEEE Trans. Automatic Control **62**(2), 753–765 (2017)
78. Mo, Y., Sinopoli, B.: Secure control against replay attacks. In: Proc. Annual Allerton Conf. Communication, Control, and Computing, pp. 91–918 (2009)
79. Momani, M., Challa, S.: Survey of trust models in different network domains. arXiv preprint arXiv:1010.0168 (2010)
80. Murguia, C., Farokhi, F., Shames, I.: Secure and private implementation of dynamic controllers using semihomomorphic encryption. IEEE Trans. Automatic Control **65**(9), 3950–3957 (2020)
81. Murguia, C., van de Wouw, N., Ruths, J.: Reachable sets of hidden CPS sensor attacks: Analysis and synthesis tools. In: Proc. IFAC World Congress, pp. 2088–2094 (2017)
82. Ng, A.: Machine Learning. Stanford University Lecture Notes (2018)
83. The NIST Privacy Framework: A Tool for Improving Privacy through Enterprise Risk Management. Technical Report (2020). Retrieved from https://www.nist.gov/privacy-framework
84. Nozari, E., Tallapragada, P., Cortés, J.: Differentially private average consensus: Obstructions, trade-offs, and optimal algorithm design. Automatica **81**, 221–231 (2017)
85. Pajic, M., Lee, I., Pappas, G.J.: Attack-resilient state estimation for noisy dynamical systems. IEEE Trans. Control of Network Systems **4**(1), 82–92 (2017)
86. Pan, Z., Baar, T.: $\mathcal{H}_\infty$ control of large scale jump linear systems via averaging and aggregation. Int. J. Control **72**(10), 866–881 (1999)
87. Pasqualetti, F., Bicchi, A., Bullo, F.: Consensus computation in unreliable networks: A system theoretic approach. IEEE Trans. Automatic Control **57**(1), 90–104 (2012)
88. Pasqualetti, F., Dorfler, F., Bullo, F.: Attack detection and identification in cyber-physical systems. IEEE Trans. Automatic Control **58**(11), 2715–2729 (2013)
89. Pasqualetti, F., Dörfler, F., Bullo, F.: A divide-and-conquer approach to distributed attack identification. In: Proc. IEEE Conf. Decision and Control, pp. 5801–5807 (2015)
90. Pawlick, J., Farhang, S., Zhu, Q.: Flip the cloud: Cyber-physical signaling games in the presence of advanced persistent threats. In: Proc. Int. Conf. Decision and Game Theory for Security, pp. 289–308. Springer (2015)
91. du Pin Calmon, F., Fawaz, N.: Privacy against statistical inference. In: Proc. Annual Allerton Conf. Communication, Control, and Computing, pp. 1401–1408 (2012)
92. Pramod, T.C., Sunitha, N.R.: Polynomial based scheme for secure SCADA operations. Procedia Technol. **21**, 474–481 (2015)
93. Sandberg, H., Teixeira, A., Johansson, K.H.: On security indices for state estimators in power networks. In: Proc. First Workshop on Secure Control Systems (2010)
94. Sanjab, A., Saad, W.: On bounded rationality in cyber-physical systems security: Game-theoretic analysis with application to smart grid protection. In: Proc. Joint Workshop on Cyber-Physical Security and Resilience in Smart Grids, pp. 1–6 (2016)
95. Sankar, L., Rajagopalan, S.R., Poor, H.V.: Utility-privacy tradeoffs in databases: An information-theoretic approach. IEEE Trans. Information Forensics and Security **8**(6), 838–852 (2013)

96. Satchidanandan, B., Kumar, P.R.: Dynamic watermarking: Active defense of networked cyber-physical systems. Proc. the IEEE **105**(2), 219–240 (2017)
97. Satchidanandan, B., Kumar, P.R.: On the operational significance of the securable subspace for partially observed linear stochastic systems. In: Proc. IEEE Conf. Decision and Control, pp. 2068–2073 (2018)
98. Shannon, C.E.: Communication theory of secrecy systems. Bell Syst. Tech. J. **28**(4), 656–715 (1949)
99. Shoukry, Y., Chong, M., Wakaiki, M., Nuzzo, P., Sangiovanni-Vincentelli, A., Seshia, S.A., Hespanha, J.P., Tabuada, P.: SMT-based observer design for cyber-physical systems under sensor attacks. ACM Trans. Cyber-Physical Systems **2**(1), 5 (2018)
100. Sun, H., Peng, C., Zhang, W., Yang, T., Wang, Z.: Security-based resilient event-triggered control of networked control systems under denial of service attacks. J. the Franklin Institute **356**(17), 10277–10295 (2019)
101. Sundaram, S., Gharesifard, B.: Secure local filtering algorithms for distributed optimization. In: Proc. IEEE Conf. Decision and Control, pp. 1871–1876 (2016)
102. Sundaram, S., Hadjicostis, C.N.: Distributed function calculation via linear iterative strategies in the presence of malicious agents. IEEE Trans. Automatic Control **56**(7), 1495–1508 (2011)
103. Tanaka, T., Skoglund, M., Sandberg, H., Johansson, K.H.: Directed information and privacy loss in cloud-based control. In: Proc. American Control Conference, pp. 1666–1672 (2017)
104. Teixeira, A., Shames, I., Sandberg, H., Johansson, K.H.: Revealing stealthy attacks in control systems. In: Proc. Annual Allerton Conf. Communication, Control, and Computing, pp. 1806–1813 (2012)
105. Teixeira, A., Shames, I., Sandberg, H., Johansson, K.H.: A secure control framework for resource-limited adversaries. Automatica **51**, 135–148 (2015)
106. Teixeira, A., Sou, K.C., Sandberg, H., Johansson, K.H.: Secure control systems: A quantitative risk management approach. IEEE Control Syst. Mag. **35**(1), 24–45 (2015)
107. Tsai, C.F., Hsu, Y.F., Lin, C.Y., Lin, W.Y.: Intrusion detection by machine learning: A review. Expert Syst. Appl. **36**(10), 11994–12000 (2009)
108. Usevitch, J., Panagou, D.: Determining r-and (r, s)-robustness of digraphs using mixed integer linear programming. Automatica **111**, 108,586 (2020)
109. Vamvoudakis, K.G., Hespanha, J.P.: Cooperative Q-learning for rejection of persistent adversarial inputs in unknown networked systems. IEEE Trans. Automatic Control **63**(4), 1018–1031 (2018)
110. Vamvoudakis, K.G., Hespanha, J.P., Sinopoli, B., Mo, Y.: Detection in adversarial environments. IEEE Trans. Automatic Control **59**(12), 3209–3223 (2014)
111. Van Dijk, M., Gentry, C., Halevi, S., Vaikuntanathan, V.: Fully homomorphic encryption over the integers. In: Proc. Int. Conf. Theory and Applications of Cryptographic Techniques, pp. 24–43 (2010)
112. Weerakkody, S., Ozel, O., Sinopoli, B.: A Bernoulli-Gaussian physical watermark for detecting integrity attacks in control systems. In: Proc. Annual Allerton Conf. Communication, Control, and Computing, pp. 966–973 (2017)
113. Weerakkody, S., Sinopoli, B.: Detecting integrity attacks on control systems using a moving target approach. In: Proc. IEEE Conf. Decision and Control, pp. 5820–5826 (2015)
114. Wiese, M., Oechtering, T.J., Johansson, K.H., Papadimitratos, P., Sandberg, H., Skoglund, M.: Secure estimation and zero-errorsecrecy capacity. IEEE Trans. Automatic Control **64**(3), 1047–1062 (2019)
115. Wiyatno, R.R., Xu, A., Dia, O., de Berker, A.: Adversarial examples in modern machine learning: A review. arXiv preprint arXiv:1911.05268 (2019)
116. Wu, N., Farokhi, F., Smith, D., Kaafar, M.: The value of collaboration in convex machine learning with differential privacy. In: Proc. IEEE Symp. Security and Privacy, pp. 485–498 (2020)
117. Zamani, M., Sadeghikhorrami, L., Safavi, A.A., Farokhi, F.: Private state estimation for cyber-physical systems using semi-homomorphic encryption. In: Proc. Int. Symp. Mathematical Theory of Networks and Systems, pp. 399–404 (2018)

118. Zhang, H., Fata, E., Sundaram, S.: A notion of robustness in complex networks. IEEE Trans. Control of Network Systems **2**(3), 310–320 (2015)
119. Zhu, Q., Baar, T.: Robust and resilient control design for cyber-physical systems with an application to power systems. In: Proc. IEEE Conf. Decision and Control, pp. 4066–4071 (2011)
120. Zhu, Q., Tembine, H., Baar, T.: Network security configurations: A nonzero-sum stochastic game approach. In: Proc. American Control Conference, pp. 1059–1064 (2010)

Part III
Cyber-Physical Security for Applications

Chapter 6
A Low-Cost Approach to Securing Commercial GPS Receivers Against Spoofing Attacks

Muaz Ahmad and Yongqiang Wang

Abstract The Global Positioning System (GPS) has been used ubiquitously for navigation and time synchronization. Nowadays, numerous telecommunication, transportation, finance, and aviation systems rely heavily on GPS for routine operations. However, due to the unencrypted nature of civilian GPS signals, commercial off-the-shelf GPS receivers are susceptible to spoofing attacks, which has been confirmed in a number of reported events. Although plenty of approaches to GPS spoofing defense have been proposed, most of these approaches require access to raw GPS signals (which are not accessible from commercial GPS receivers) or additional navigation sensors (which increase cost or engineering time), making them inappropriate for mass-market applications to secure commercial GPS receivers. We propose a low-cost spoofing detection approach for vehicular GPS receivers by purposely perturbing and then examining changes in the Doppler shift that are directly reported by most commercial GPS receivers. More specifically, by purposely enforcing vehicle velocity variations that are unpredictable to potential spoofers, we can enforce unpredictable variations to the actual Doppler shift on a GPS receiver. Then, through cross-checking if the Doppler shift value reported by a GPS receiver is consistent with the variation in velocity, we can tell if the GPS receiver is receiving genuine GPS signals or not. The variation in vehicle velocity is directly accessed from the vehicle On-Board Diagnostics (OBD) system available on all cars manufactured after 1996, which avoids using extra navigation sensors. The proposed approach requires neither dedicated radios to acquire raw GPS signals nor extra navigation sensors and is directly applicable to commercial off-the-shelf receivers, making it promising for mass-market applications to protect commercial GPS receivers. Extensive field experiments were conducted to systematically evaluate the performance as well as potential limitations of the proposed approach.

M. Ahmad · Y. Wang (✉)
Clemson University, Clemson, South Carolina 29634, USA
e-mail: yongqiw@clemson.edu

H. Ishii and Q. Zhu (eds.), *Security and Resilience of Control Systems*,
Lecture Notes in Control and Information Sciences 489,
https://doi.org/10.1007/978-3-030-83236-0_6

6.1 Introduction

The Global Positioning System (GPS) was originally introduced by the US Department of Defense for military applications. Since allowed for civilian usage in the 1980s, particularly following the discontinuation of the "selective availability" feature in the year 2000, it has become an essential component for numerous civilian applications, and its usage is expected to exponentially rise in the future with the proliferation of Internet of Things (IoT) devices using localization services. Nowadays besides the commonly known application of positioning and navigation, GPS is also widely used for precise time synchronization and frequency calibration. For example, most cell-phone towers use GPS signals to calibrate the frequency and timing of wireless transmissions. Power systems use GPS to precisely time stamp and correlate voltage and current measurements conducted at distributed Phasor Measurement Units. GPS is also utilized in time stamping security videos and synchronizing traffic light timing [1].

The heavy reliance on civilian GPS also makes the security of GPS an important problem that is receiving increased attention. The civilian GPS signal is completely unencrypted with no authentication, which makes it vulnerable to attacks. Compared with attacks trying to jam GPS signals, GPS spoofing attacks are more dangerous as they can arbitrarily change the time/position of a GPS receiver without being noticed. Using the publicly known spreading codes for GPS satellites, modulation schemes, and data structure, a GPS spoofer mounts spoofing attacks by broadcasting authentic-appearing GPS signals with valid modulation codes, signal structure, and frequency but fakes information to trick a victim receiver into misinterpreting them as authentic signals and hence deduces a false position fix, a false clock offset, or both. In [1], the authors proved that such spoofing attacks can completely fool commercial off-the-shelf GPS receivers in use today by arbitrarily manipulating the time and position reporting at the will of the spoofer. Moreover, according to [2], a team of researchers was able to show that spoofing attacks were effective in diverting the path of an $80 million private yacht by seamlessly taking over its GPS signal. There have also been unconfirmed claims of GPS spoofing being used to sabotage drones or explain cases of mysterious missing planes and ships [3]. With the ubiquitous use of GPS-based services, which are only expected to grow further, the threat of GPS spoofing will only grow. Besides modifying the course of ships or drones to insecure regions, changing the timing of automated stock exchange events, or corrupting the time stamp of smart grid measurements, a GPS spoofer could even cause accidents and unnecessary traffic congestions by altering emergency routes. For example, by manipulating vehicle positions at intersections, vehicles could make wrong turns or run into oncoming traffic. Such attacks become particularly probable with the advent and proliferation of (semi)autonomous vehicles which completely rely on GPS for navigation.

The vulnerability of GPS has motivated plenty of efforts to counter spoofing attacks against civilian GPS receivers. Existing reported efforts are mainly based on one of the following mechanisms: (1) using extra navigation sensors; (2) employing signal encryption; (3) using multiple GPS receivers (formation) or multiple antennas (array); and (4) processing of raw GPS signals. Detailed explanations of these mechanisms are given in Sect. 6.2. Although effective, these existing efforts either rely on specialized hardware such as software-defined radios to acquire/process raw GPS signals which impose modifications on existing commercial off-the-shelf receivers or entail extra navigation sensors for consistency checking which severely increases complexity, cost, or engineering time. (Note that approaches based on extra navigation sensors are also limited in their performance because extra navigation sensors like IMU are subject to errors that accumulate quickly due to inherent biases [4].) Therefore, existing GPS spoofing countermeasures are limited in their applicability to commercial off-the-shelf GPS receivers and hence are unsuitable for mass-market deployment. In fact, as corroborated by a comment made in a recent paper [5], "existing countermeasures are unlikely to be implemented in the near future since they all require far-reaching modifications of either the GPS infrastructure or the receiving devices".

Driven by the increasing threat on commercial GPS receivers but the lack of effective security solutions, we propose a security approach to protect vehicular GPS receivers against spoofing attacks by purposely perturbing and then cross-checking the corresponding response in the Doppler shift that is directly reported by most commercial off-the-shelf GPS receivers. More specifically, by purposely enforcing vehicle velocity variations that are unpredictable to potential GPS spoofers (we assume that GPS spoofers are outside but not in a target vehicle), we can enforce variations to a GPS receiver's Doppler shift report that is unpredictable to potential spoofers. Then through cross-checking if the reported Doppler shift is consistent with the imposed variation in velocity, we can tell if a GPS receiver is receiving genuine GPS signals or not. It is worth noting that we access the variation in vehicle velocity directly from the vehicle On-Board Diagnostics (OBD) system available on all cars manufactured after 1996, which avoids imposing extra navigation sensors. It is also worth noting that the required *speed variations can be performed intentionally for spoofing detection purposes when driving conditions permit. They can also be natural maneuvers that drivers have to make according to road conditions when lane changing, speeding up, slowing down, overtaking, or turning are performed.* The avoidance of extra specialized radios or navigation sensors makes the approach directly applicable to commercial off-the-shelf GPS receivers with the assistance of limited general-purpose computing which can be fulfilled by a smartphone or a Raspberry pi board. Combined with the additional merit of being implementable by individual commercial GPS users in a completely autonomous way, the proposed approach is extremely promising for mass-market applications to protect commercial off-the-shelf GPS receivers.

The chapter is organized as follows. Section 6.2 reviews relevant literature. Section 6.3 introduces the background of the research, including the basic principles of GPS. Section 6.4 introduces the attack model, which is followed by our security approach in Sect. 6.5. Experimental validation results of the security approach are presented in Sect. 6.6, which is followed by conclusions in Sect. 6.7.

6.2 Literature Review and Contributions

Since the publication of the Volpe report in 2001 warning the potential threat of GPS spoofing [6], plenty of efforts have been reported to defend against such attacks. Some of the typical defending approaches for GPS receivers are summarized in Table 6.1.

The first class of spoofing-defending approaches is based on signal encryption. Such approaches propose to authenticate signals from satellites with unpredictable signatures that are only usable by legitimate users with a secret key. For example, the authors in [10–12] proposed to insert public-encryption-based digital signature into navigation messages. Alternatively, the authors in [10, 13] proposed to interleave spectrum security code with normal civil GPS spreading codes to make the GPS spreading sequence periodically unpredictable. However, such encryption-based approaches require significant modifications to the signal structure of the GPS system, which cannot be implemented without significant modifications to the underlying infrastructure in use by the legacy GPS system. Furthermore, such approaches do not protect against signal replaying attacks [17].

The second class of reported GPS spoofing-defending approaches relies on the acquisition and processing of raw GPS signals. Such approaches determine the authenticity of GPS signals by checking if expected code signatures of the raw signals are present. For example, the authors in [1] proposed to correlate the raw-signal sampling of the encrypted P(Y) code between a potentially vulnerable GPS receiver and a secure reference receiver. Because the time-varying P(Y) code is unknown, and hence cannot be replicated by the spoofer in real time, a high correlation value between the secure reference receiver and the vulnerable receiver can only be obtained when the vulnerable GPS receiver is receiving genuine GPS signals from GPS satellites as the reference receiver does. The authors in [8] expanded on the concept of P(Y) code correlation and proposed to perform P(Y) code correlation across multiple cooperative receivers, which can improve the probability of detecting a spoofing attack. The authors in [7] proposed to detect traces of the authentic

Table 6.1 Commonly used GPS spoofing detection approaches and their limitations

Approaches based on	No specialized hardware	Easy to scale	Representative papers
Processing of raw GPS signals	X	✓	[1, 7]
Antenna array or multi-receivers	✓	X	[8, 9]
Encryption of GPS signals	X	X	[10–13]
Extra navigation sensors	X	✓	[14, 16]
Approach in this chapter	✓	✓	

GPS signal in the presence of fake spoofed GPS signals. If the spoofed signal is not closely synchronized with the authentic GPS signal, there would be the presence of shifted and overshadowed auxiliary peaks that would indicate the presence of spoofing. All of these approaches require access to the raw GPS signals and hence rely on software-defined radios to sample and process high-frequency acquisition-phase GPS signals. Because such raw GPS signals are not accessible from existing commercial off-the-shelf GPS receivers, the above approaches face significant hurdles in mass-market deployment to protect commercial off-the-shelf GPS receivers.

The third class of reported spoofing-defending approaches employs extra navigation sensors such as Inertial Measurement Unit (IMU), magnetometers, and accelerometers to cross-check if information reported by a GPS receiver is consistent with the reporting of the extra navigation sensor. Such approaches work under the assumption that the extra navigation sensor cannot be manipulated by a spoofer. For example, by using an extra magnetometer-disciplined inertial navigation system, the authors in [14] proposed to counter GPS spoofing by checking if GPS reporting is consistent with the information provided by the extra inertial navigation system. Advantages of such a defending mechanism are that extra navigator sensors such as IMU are hard for a GPS spoofer to manipulate, which guarantees a certain level of protection strength. Disadvantages of such approaches are also clear in that the requirement of extra navigation sensors increases cost, complexity, engineering time, or power requirements. Furthermore, because it is very hard for the extra navigator system to provide position and velocity with similar accuracy and long-term stability as the GPS-based navigation (e.g., IMU velocity and position measurements are subject to errors that accumulate quickly due to IMU biases [4]), such approaches cannot detect spoofers using small deviations to deceive GPS receivers.

Other approaches that do not require specialized receivers or sensors have also been proposed. For example, the authors in [9] proposed to use an antenna array to distinguish if received signals are from GPS satellites or not. As the authentic GPS signals received by a GPS receiver are broadcasted by different satellites with different positions in the constellation, codes from different satellites will experience different relative phases across antennas in the antenna array. On the contrary, a GPS spoofer usually broadcasts from a single source, and all the codes would have the same relative code phase across the antennas, which could indicate spoofing. The approach is implementable in a static case and requires a collection of strategically placed antennas, which limits its scalability potential. In [18], the authors proposed to use multiple GPS receivers placed in a formation to counter spoofing attacks. This approach limits the positions from which a spoofer can succeed in spoofing a single GPS receiver without disturbing the established formation. Such an approach requires multiple GPS receivers deployed in a known static formation with substantial relative distances, which is costly and restricted by practical deployment conditions.

More simplistic detection techniques have also been discussed for detecting GPS spoofing based on consistency checks. They verify the expected GPS properties and ensure that signal properties such as the signal-to-noise ratio, signal correlation quality, direction of arrival of the signal, and code/phase rates are in the expected range [8]. These consistency checks, however, are easy to circumvent by a sophisticated spoofer in its fake signal generation and hence can only detect rudimentary GPS spoofers. In this chapter, we ignore these weak verification mechanisms and only focus on countermeasures that can defeat sophisticated spoofers.

Table 6.1 summarizes some of the mostly used approaches for GPS spoofing detection and their limitations. It can be seen that existing approaches either need specialized hardware (specialized GPS receivers or extra navigation sensors) or are difficult to scale, limiting their potential mass-market deployment to protect commercial off-the-shelf GPS receivers.

In this chapter, we propose a GPS spoofing detection approach based on purposely perturbing and then cross-checking corresponding response in the Doppler shift of received signals. The Doppler shift is a change in the nominal signal frequency (1.5424 GHz GPS signals emitted by GPS satellites) due to the relative motion between a GPS receiver and GPS satellites. Most commercial GPS receivers report the Doppler shift in frequency, with some examples given in Table 6.2. We access the Doppler shift reported in commercial GPS receivers and then correlate it with the calculated Doppler shift at the receiver based on the satellite and receiver relative motion to detect potential inconsistencies caused by spoofing attacks. It is worth noting that Doppler shift-based GPS signal verification has been discussed in [19], which uses satellite motion to predict the Doppler shift and determine potential inconsistencies. However, as the motion of the satellites can be predicted by the spoofer, the spoofer can adjust the output frequency of its fake signal accordingly to outflank detection. In this chapter, to avoid the spoofer from precisely predicting the Doppler shift in receiver's signals based on known satellite motion, we enforce unpredictable, yet monitored velocity variations on the GPS receiver, which makes the actual Doppler shift value impossible for a spoofer to predict precisely. To monitor velocity variation of the receiver, we make use of the digital vehicle speed report from the OBD system, which is installed in all cars manufactured after 1996. With reliable velocity variation values and the satellite motion, the actual Doppler shift value of a receiver's received signals can be computed in real time and used to cross-check with the Doppler shift reported by commercial GPS receivers. If there is inconsistency, then it is safe to conclude the occurrence of spoofing attacks on the receiver. Unlike most security approaches, our approach does not require special raw GPS signal acquisition/processing radios such as software-defined radios nor does our approach need an array of antennas or a formation of receivers. These properties make the

Table 6.2 Commercial GPS receivers reporting the Doppler shift

Brand	Device	Cost
Ublox	NEO-M8T	$75 [20]
SkyTraq	NS-RAW	$70 [21]
NVS	RasPiGNSS	$170 [22]
Swift	Piksi Multi GNSS Module	$595 [24]
NovAtel	OEM625S	Unknown

proposed approach promising for mass-market deployment to protect commercial GPS receivers against spoofing attacks. It is worth noting that the spoofing-attack detection approach has been extended to spoofing-attack localization for vehicular GPS receivers in [15].

6.3 Background

6.3.1 GPS Operation Principle

GPS receivers get position and time solution by using the concept of trilateration. For accurate navigation, a GPS receiver needs to compute its distances relative to at least four GPS satellites in orbit, whose positions can be obtained by using the ephemeris extracted from received GPS signals. Four satellites are needed because there are four unknowns that have to be solved simultaneously: x, y, and z coordinates of the receiver and the GPS time, as described in Fig. 6.1.

Represent the coordinate of satellite S_i as $[x_i^s, y_i^s, z_i^s]$ and its distance to the receiver as D_i. Also represent the receiver time difference with respect to the GPS time as δ, which results in a range measurement error as $\Delta = c\delta$ with c denoting the speed of light. Then a GPS receiver uses four simultaneous equations below to solve for the receiver's four unknowns:

$$\begin{cases} (D_1 - \Delta)^2 = (x - x_1^s)^2 + (y - y_1^s)^2 + (z - z_1^s)^2 \\ (D_2 - \Delta)^2 = (x - x_2^s)^2 + (y - y_2^s)^2 + (z - z_2^s)^2 \\ (D_3 - \Delta)^2 = (x - x_3^s)^2 + (y - y_3^s)^2 + (z - z_3^s)^2 \\ (D_4 - \Delta)^2 = (x - x_4^s)^2 + (y - y_4^s)^2 + (z - z_4^s)^2 \end{cases}. \tag{6.1}$$

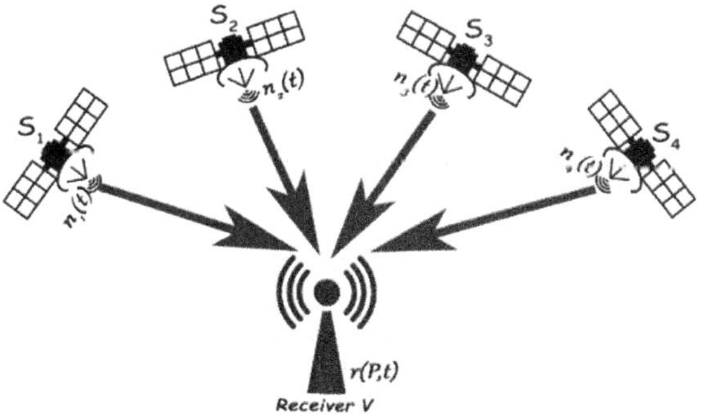

Fig. 6.1 The principle of satellite positioning. A receiver V uses signals from at least four satellites $n_i(t)$ to find its position and time $r(P, t)$

6.3.2 GPS Receivers

The main objective of the GPS receiver is to extract the pseudorandom codes and the navigation messages from the complex signals received from GPS satellites. GPS receivers operate by passing received signals through four main processing blocks. A brief description of the processing and extraction of GPS signals is given below. More details can be found in [25].

6.3.2.1 RF Front End

The front end comprises the antenna module receiving the GPS signal and an Analog-to-Digital Converter (ADC), which converts the signal to the digital form. The antenna receives GPS signals from the electromagnetic spectrum and converts it into electrical currents and voltages. Due to the weak strength of the GPS signal, it is amplified with a Low-Noise Amplifier (LNA) unit, and in most cases, the carrier frequency is down converted to an intermediate frequency which is easier to process with lower stress on the computational hardware [26].

6.3.2.2 Acquisition Module

The acquisition module is responsible for detecting the presence of satellite data in the received signal. It does so by making use of a Delay Locked Loop (DLL) to properly align the pseudorandom codes (C/A codes) with a local replica generated by the receiver [27]. Once the preprocessing is complete, the acquisition module performs a scan for satellite pseudorandom codes. The scan is over a frequency range of 10 kHz [27] around the GPS satellite broadcasting frequency to ensure that all the possible Doppler frequency shifts are accounted for. The DLL constantly aligns the replica codes with the incoming data for correlation beyond an acceptable threshold. The C/A codes from different satellites are perpendicular to each other, thus there would be a very low correlation between non-aligned codes. With the satellites in orbital motion around the receiver, there is the presence of a frequency shift known as the Doppler shift due to which the frequency received by the receiver is shifted from the satellite broadcasting frequency. Thus, a two-dimensional search is performed in this phase to search for the Doppler shifts and C/A codes.

6.3.2.3 Tracking Module

The tracking module keeps track of the code phase and the frequency as the signal changes with time. The code phase is tracked in a DLL just like in the acquisition phase. However, here the corresponding pseudorandom code is already known. It realigns the signal if it has been shifted by realigning it to closely shifted replicas of the already known pseudorandom code. The replicas are normally shifted by about a half chip. The tracking phase operates on the signal in real time and continuously updates to track all changes. If the receiver loses track of a satellite, the cycle must back track to the acquisition phase and reacquire the satellite to proceed [26].

6.3.2.4 Position, Velocity, and Time (PVT) Module

The extracted navigation message includes the signal transmission time, which is accurately generated by the satellite. To estimate the user position, the receiver first estimates the satellite position at the time of transmission using the ephemeris included in the signal. Secondly, the signal transmission and reception times are used to estimate the receiver's relative distance to a satellite. It should be noted that the received time is based on the receiver's clock which could be subject to errors. Therefore, the estimated range is referred to as the pseudorange. Pseudoranges from at least four satellites are needed to estimate the receiver's position and time by using Eq. (6.1).

6.4 Attack Model

A spoofing attack can deceive GPS receivers into calculating false position and timing solutions. Since civilian GPS signals are publicly known, a spoofer can generate authentic-appearing GPS signals. Due to the distance of the GPS satellites, the signals processed by a receiver are very weak, which makes it easy for a spoofer to broadcast stronger manipulative signals that could overshadow the authentic signals and further take control of GPS receivers. We assume that GPS spoofers are outside but not in a target vehicle. In this chapter, we consider spoofing attacks mounted using the publicly accessible open-source spoofing library GPS-SDR-SIM [28], which can create digital GPS signals with custom position and time. The GPS-SDR-SIM library is used in almost all existing GPS spoofing studies. Below we describe how a spoofer incorporates wrong information and generates fake signals. It is worth noting that as vehicles purposely enforce velocity variations that are unpredictable to potential spoofers, even if a spoofer has the capability to measure the target vehicle's velocity/direction of traveling, calculate Doppler frequencies, and adjust its transmission frequency accordingly, there is always latency in this process. For example, for the spoofer to measure the velocity and direction of traveling of a target vehicle (both are necessary to find the Doppler effect), there will be inevitable delays due to the time-delayed calibration of the two measurements from different sensors [29]. There will also be delays in generating and emitting fake signals at the right (time-varying) frequency. Furthermore, commercial-grade GPS receivers are also subject to at least several seconds latency in acquiring a fix when message codes/contents are changed (for the purpose of spoofing) [30]. These inevitable delays are important for our detection mechanism, which are experimentally evaluated in detail later in Sect. 6.6.3.

6.4.1 Code Level Attacks

The C/A codes are received by a GPS receiver in the acquisition phase using DLL. As described earlier, this process repeatedly aligns the incoming collections of C/A codes with the corresponding local replica to keep track of individual satellites. Once a satellite is identified, the shift in the correlation between the incoming signal and the local replica is used to calculate the pseudorange. Because a spoofer can manipulate the correlation shift by changing its emitted fake signal, the pseudoranges obtained by a GPS receiver can be manipulated by the spoofer at will. On receiving such a fake signal, the receiver would obtain wrong pseudoranges to available satellites, which will lead to incorrect position and/or time solutions when used in the equations in (6.1).

6.4.2 Navigation Message Attacks

The navigation message contains satellite information such as the GPS time and the ephemeris. The ephemeris provides satellite and orbital parameters that are used to compute the position of satellites in orbit. A spoofer can create fake GPS signals and meddle with the ephemeris variables and/or GPS time values in the navigation message to mislead the receiver with false coordinates. These false coordinates will cause the receiver to solve for false positional information as well. Figure 6.2 shows a schematic of how tampering with the navigation message can cause the receiver to obtain false satellite and user positions.

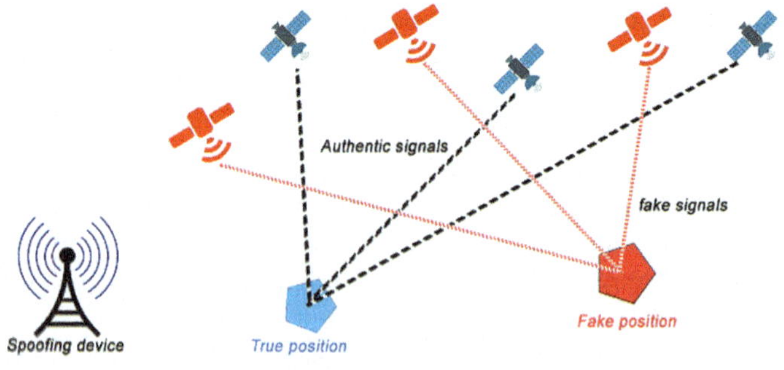

Fig. 6.2 GPS Spoofing. Corrupted ephemeris can cause the receiver to position itself relative to wrongly positioned satellites and in turn resolve a wrong position/time solution

6.4.3 Experimental Confirmation of Spoofing-Attack Models

Abiding by legal constraints, we experimentally implemented GPS spoofing attacks according to the described attack models. We used a software-defined radio as the front end to broadcast signals at the GPS nominal frequency. Two software-defined radios USRP from National Instruments™ and HackRF from Great Scott Gadgets™ were used in our tests. To demonstrate the effectiveness of easily implementable and portable spoofing devices, this chapter only discusses results based on the HackRF. We used the GPS-SDR-SIM library to create binaries with valid GPS navigation messages and C/A codes that can be broadcasted by the HackRF at the GPS nominal frequency of 1.5745 GHz. We used the NEO-M8T Ublox GPS chip [20] as our test receiver. It gives access to GPS observables such as the code phase, the Doppler shift, and the navigation message. The setup successfully spoofed receiver location to an arbitrary point on the globe under both code-level and navigation message attacks. It is worth noting that because interfering with GPS signals in open uncontrolled space is illegal [1], we directly wired the HackRF output to the GPS receiver (bypassing the antenna) to ensure that all our tests did not affect nearby GPS receivers. Furthermore, to minimize possible emanated electromagnetic signals, the HackRF and used GPS receiver were boxed-in and wrapped with aluminum foil to shield any signal from leakage and ensure that there was no interference with any nearby receivers. The spoofing setup is illustrated in Fig. 6.3.

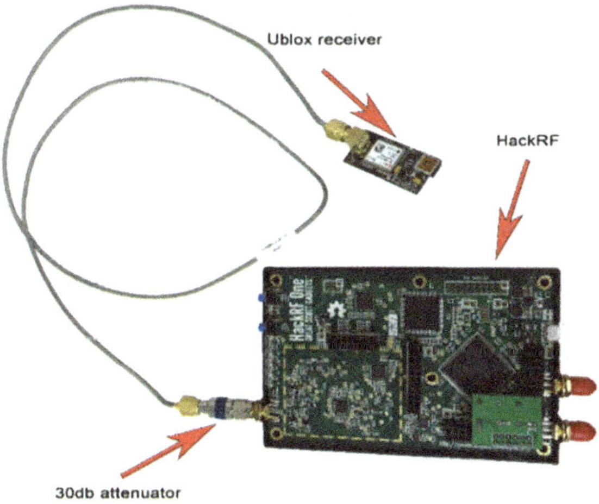

Fig. 6.3 The experimental setup where the HackRF was directly connected to the Ublox GPS receiver

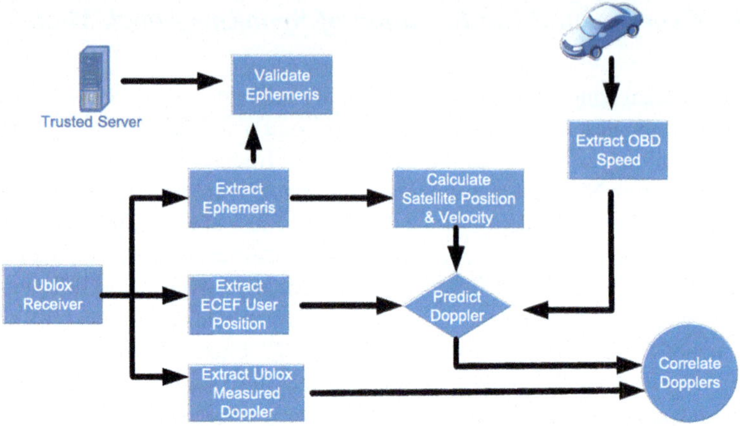

Fig. 6.4 Overview of the GPS security approach

6.5 Security Approach

A schematic diagram of our security approach against GPS spoofing is shown in Fig. 6.4. Our approach first validates the GPS navigation messages to ensure that they are valid. It then compares the reported Doppler shift with the calculated counterpart to detect possible inconsistencies caused by spoofing attacks. However, it is worth noting that the first step of validating navigation messages is just a backup detection mechanism and can be removed if practical conditions do not permit (e.g., connection to a server is impossible). This is because our Doppler-based detection mechanism can catch both types of attacks, neither of which can generate a fake signal with the expected actual Doppler shift.

6.5.1 Ephemeris Validation

We set up a secure server to check if the navigation message has been tampered with by a spoofer. A reliable GPS receiver at our university is used to extract real-time ephemeris variables from navigation messages for each satellite. The extracted ephemeris variables are pushed to a Github repository, which is accessible to target GPS receivers. A target GPS receiver intermittently pulls the ephemeris from the server and correlates the trusted ephemeris with the local counterpart to check if the ephemeris has been tampered with. As mentioned earlier, this step of ephemeris validation can be removed when access to a server is impossible.

6.5.2 Doppler Frequency Validation (Static Receiver Case)

The Doppler shift can be calculated accurately by using the motion vectors of GPS satellites with respect to the receiver. The following equation lets us find the GPS Doppler shift under the assumption that the clock error is almost constant [31]:

$$D = F_t - F_R, \tag{6.2}$$

$$F_r = F_t(1 - V_R/c), \tag{6.3}$$

where F_t is the GPS broadcasting frequency, F_R is the received frequency, and V_R is the relative velocity vector between the satellite and the receiver along their line of sight. The relative velocity is given as follows:

$$V_R = (V_S - V_r) \times (P_r - P_S)/|P_r - P_S|, \tag{6.4}$$

where V_s is the satellite velocity, V_r is the receiver velocity, P_s is the satellite position, and P_r is the receiver position. The satellite velocity and position are calculated from the ephemeris based on a set of standard celestial physics equations described in [27]. The receiver position is obtained from GPS output.

Having access to the Doppler shift measurement from the Ublox receiver and the expected Doppler shift using (6.2)–(6.4), we can make a comparison to detect potential spoofers unable to incorporate the Doppler shift into spoofed signals. Figures 6.5 and 6.6 illustrate the calculated and reported Doppler shift values in our test using the Ublox GPS chip in the absence of spoofing attacks. The measured and calculated Doppler shift values are only off by a few hertz at most. Therefore, we can indeed use (6.2)–(6.4) to have the Doppler shift and compare it against the reported Doppler shift value to detect possible spoofers that cannot adjust their spoofing frequencies to incorporate the influence of the Doppler shifts in its spoofing signals. However, as mentioned earlier, the Doppler shift can be predicted by the spoofer as well. Therefore, if a spoofer does predict the Doppler shift using accessible GPS satellite velocity information and incorporates it into its fake signal generator, it can easily circumvent such a consistency-based detection mechanism.

6.5.3 Doppler Frequency Validation (Mobile Receiver)

To combat GPS spoofers that can accommodate the Doppler shifts in generated spoofing signals, we leverage the flexibility in user velocity that can also affect the Doppler shift values. More specifically, we propose to at times enforce velocity variations unpredictable to the spoofer, which will make the actual Doppler shift unpredictable by the spoofer. Of course, the velocity variation should be accessible to the user vehicle itself to find the precise Doppler shift for comparison with the receiver's reported Doppler shift. If the GPS receiver is tracking genuine GPS signals,

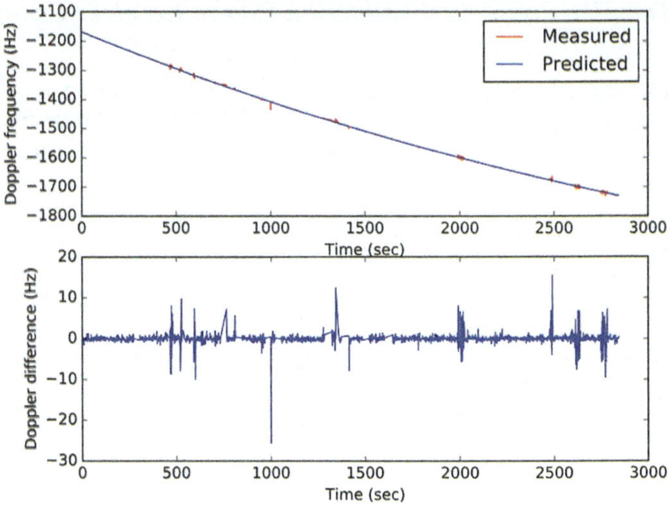

Fig. 6.5 Measured and predicted Doppler shifts for satellite no. 6 under a static receiver

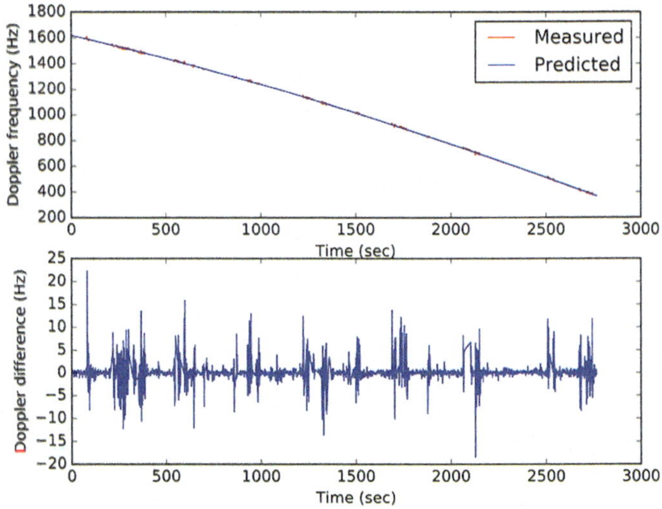

Fig. 6.6 Measured and predicted Doppler shifts for satellite no. 9 under a static receiver

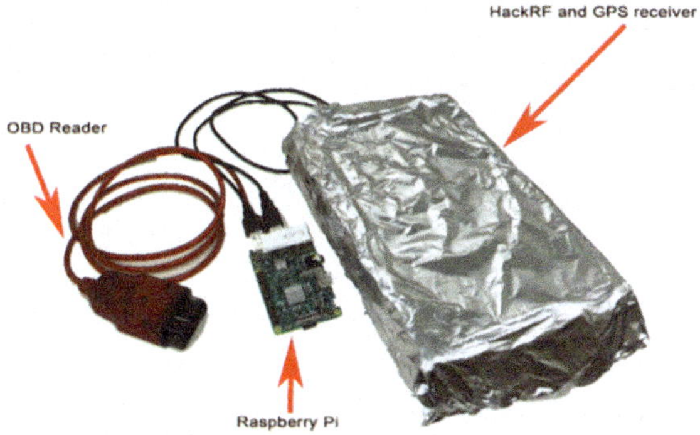

Fig. 6.7 The complete setup used in field evaluation

then the user-calculated and receiver-reported Doppler shift values will be consistent. However, if the receiver is spoofed, then its reported Doppler shift will reflect the Doppler shift value used by the spoofer in its generation of spoofing signals, which will be different from the actual Doppler shift.

To obtain the user (vehicle) velocity, some literature uses Inertial Measurement Units (IMU). However, this approach accumulates error very quickly, and the velocity estimates are unusable within a few seconds. Most of the existing digital speedometers and mobile applications utilize GPS for their speed estimation, which would be rendered useless under a spoofing attack on GPS signals. We propose to use an OBD reader to acquire accurate digital speed from vehicles. All vehicles manufactured in the U.S. after 1996 have an OBD port. This feature enables vehicles to report diagnostic and sensory information to vehicle owners and/or repair technicians. The OBD reader we used is an off-the-shelf cable that can acquire the vehicle odometer reading (cf. Fig. 6.7). It enables us to precisely monitor the enforced velocity variation that is unknown to the spoofer and hence to calculate the actual Doppler shift independently of GPS-related sources.

We experimentally evaluated if it is possible to precisely predict the Doppler shift when the receiver is in motion. We tested our algorithm (6.2)–(6.4) using a Ublox GPS chip placed in a 2008 Honda Civic. The car velocity varied between 0 and 80 km/h in the evaluation. The calculated and measured Doppler shift values are given in Figs. 6.8 and 6.9. (It is worth noting that the satellite ID changed as visible satellites constantly change with time.) It can be seen that the errors between measurement and calculation are slightly larger than in the static case.

We noticed that vehicle motions made the reported Doppler shift value very noisy. To suppress such measurement noise, we designed a Kalman filter to smooth the Doppler shift reported by the Ublox GPS chip. We choose the Kalman filter due to its ability to operate on very little system data without much knowledge of the system model. We use a simple one-dimensional filter given below:

$$\hat{x}_k = (1 - K_k)\hat{x}_{k-1} + K_k z_k, \tag{6.5}$$

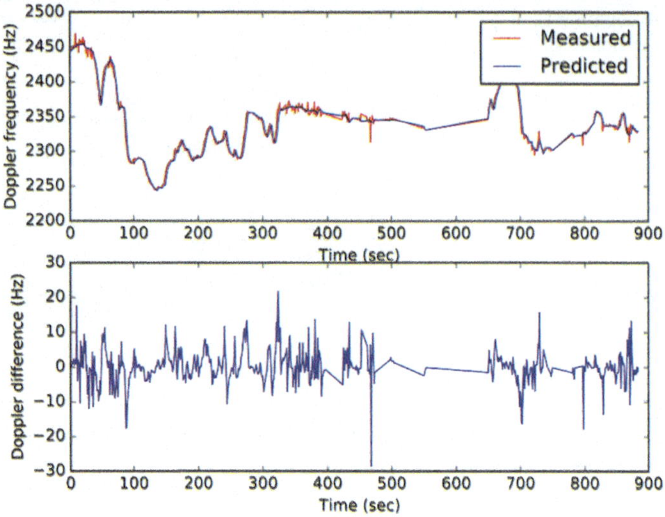

Fig. 6.8 Measured and calculated Doppler shifts for satellite no. 10 when the GPS receiver was in motion

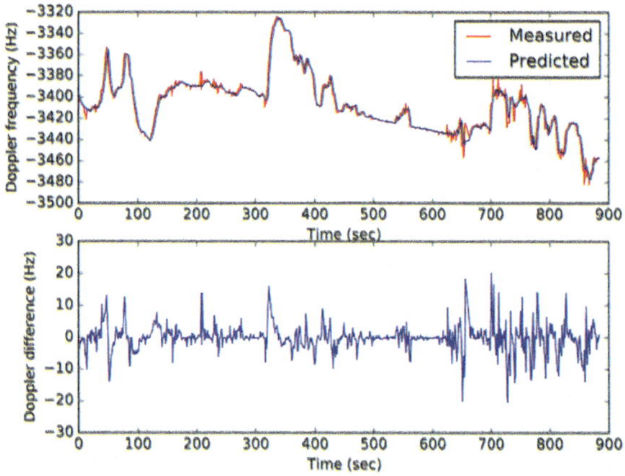

Fig. 6.9 Measured and calculated Doppler shifts for satellite no. 29 when the GPS receiver was in motion

where $\hat{x}_k$ and z_k represent the filtered and original measurements, respectively, and K_k is the Kalman gain designed according to [32].

To confirm the effectiveness of our Doppler shift calculation algorithm (6.2)–(6.4) under different environmental conditions that could potentially affect the signal propagation and further the Doppler shift, we conducted experiments at different times of day and under different weather conditions. The satellite with the largest prediction error after applying the Kalman filter was shown in Figs. 6.10, 6.11, 6.12, and 6.13. The error between the measured and calculated Doppler shift values was constrained within 10 Hz, except in the cloudy weather, where the error reached 20 Hz

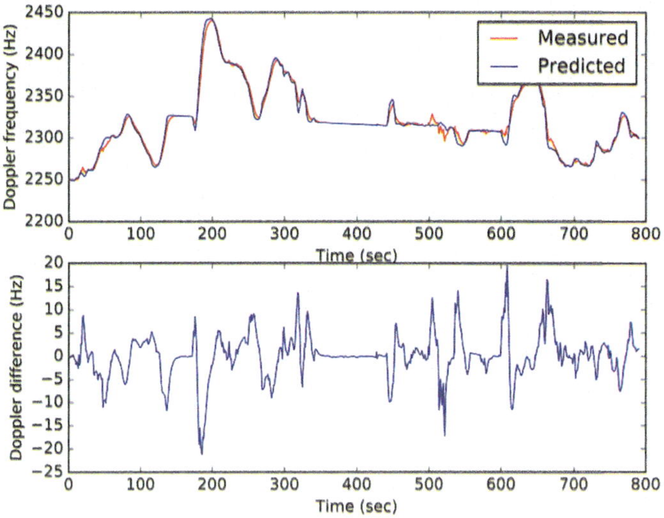

Fig. 6.10 Measured and calculated Doppler shift values at 9 am with a cloudy sky

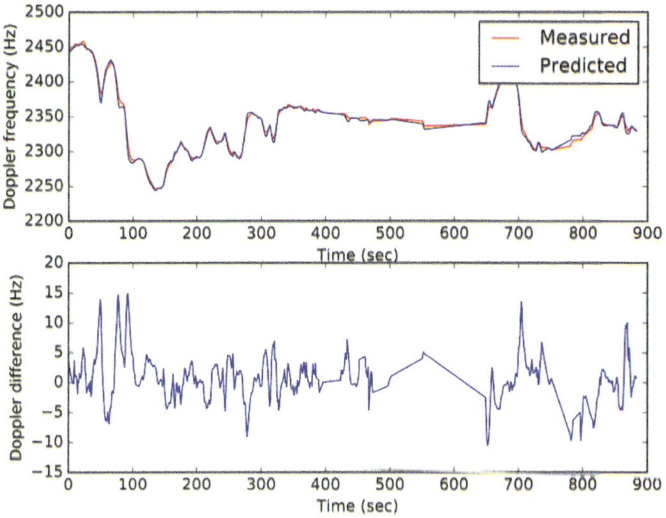

Fig. 6.11 Measured and calculated Doppler shift values at 1 pm with a clear sky

at times. However, this deviation was for the satellite with the largest prediction error. To get a more statistically reasonable result, we utilized the mean of the three satellites that have the three highest prediction errors. Such an average was also used later on for all spoofing-attack cases.

In the case with a spoofer, the user can still use the vehicle velocity from the OBD reader to find the Doppler shift. However, the spoofer was unable to access the

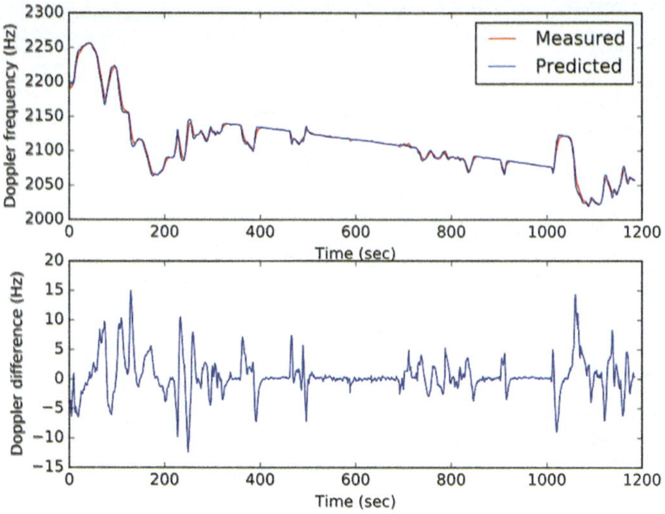

Fig. 6.12 Measured and calculated Doppler shift values at 12 pm with a clear sky

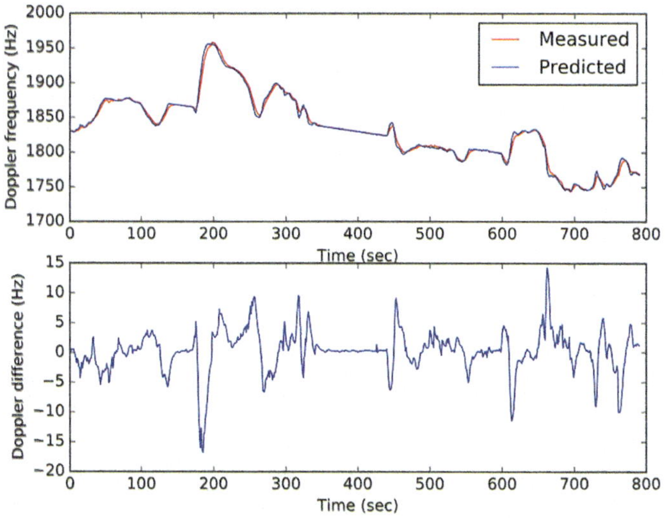

Fig. 6.13 Measured and calculated Doppler shift values at 9 pm with a cloudy sky

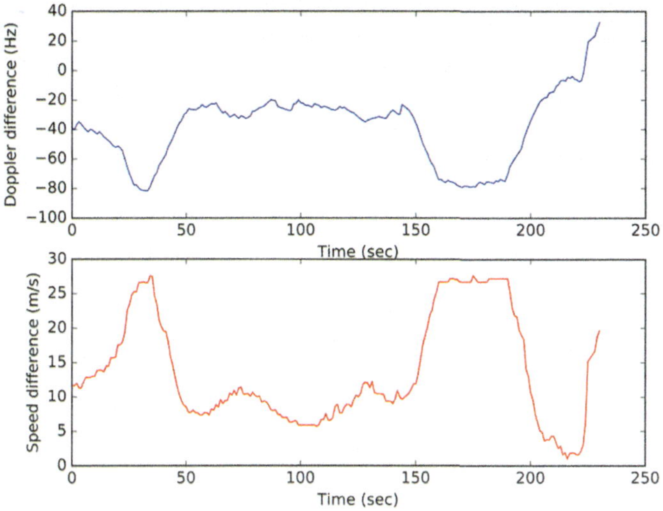

Fig. 6.14 Difference in the measured and calculated Doppler shift values for satellite 20 under different receiver velocities

velocity changes when purposed velocity variations were enforced, so there should be inconstancy in the user-calculated and receiver-reported Doppler shift values if the receiver is tracking fake signals generated by the spoofer. Figures 6.14 and 6.15 show the difference in the calculated and measured Doppler shift values in real-time spoofing tests where the GPS receiver was spoofed by the attacker. The spoofer assumed a constant velocity in generating its spoofing signals while the user Doppler computation was based on the speed acquired from the moving vehicle through the OBD reader. From Figs. 6.14 and 6.15, it could be seen that there is a significant difference between the calculated and measured Doppler shift values due to the spoofer's inability to acquire the actual vehicle velocity/direction of traveling and incorporate corresponding Doppler shift in its signal generation. Furthermore, the two figures also showed that with an increase in the difference between the actual user velocity and spoofer assumed velocity, the difference in the Doppler shift values increases, which is intuitive. It is worth noting that the Doppler differences vary between different satellites because the receiver has different relative motions with respect to different GPS satellites. Given that GPS satellites visible to the receiver keep changing with time, we searched for the four satellites with the four largest Doppler differences and used their average value to measure the Doppler difference in our field test in the following section.

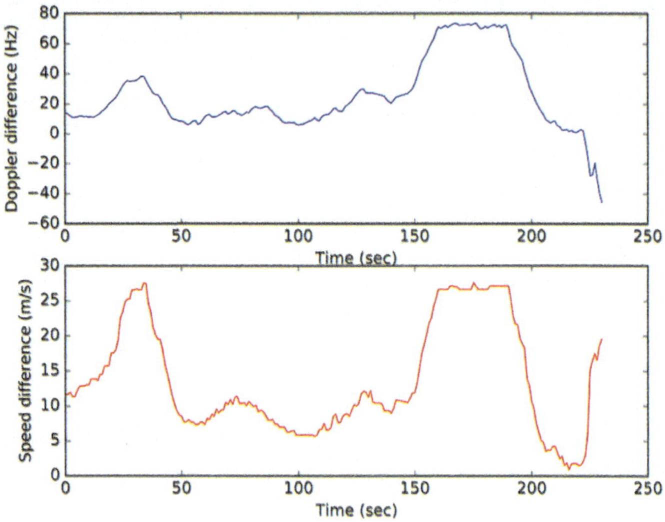

Fig. 6.15 Difference in the measured and calculated Doppler shift values for satellite 32 under different receiver velocities

6.6 Results and Analysis

We conducted a series of field tests to evaluate the performance of the proposed GPS spoofing detection approach. We also evaluated the potential limitation of our approach in combating spoofers using the GPS-SDR-SIM open-source spoofing library. As indicated earlier, to our knowledge, almost all existing GPS spoofing studies are based on the GPS-SDR-SIM library.

6.6.1 Spoofing Detection Rate Analysis

To evaluate the detection rate of the proposed spoofing detection approach, we set up the spoofer as described in Sect. 6.4 with a specified target route and velocity. The spoofer injected spoofing GPS signals to a Ublox GPS chip placed on a 2008 Honda Civic. Experimental results confirmed that the receiver was indeed spoofed as its reported position did correspond to the direction and motion of the spoofed route. The Doppler calculation was performed using the velocity accessed through the OBD reader. We conducted tests using five different Ublox GPS chips on over 20 routes and under different discrepancies between spoofer's estimated velocity and the actual velocity to evaluate the performance of our approach. Two routes were used in the test: The East–West route is a 4.3 mi route on High Way 123 (between 13330 Clemson Blvd, Seneca, SC 29678 and 10121 Clemson Blvd Suite F, Seneca, SC 29678) and the North–South route is a 8.2 mi route on High Way 133 (between 5200

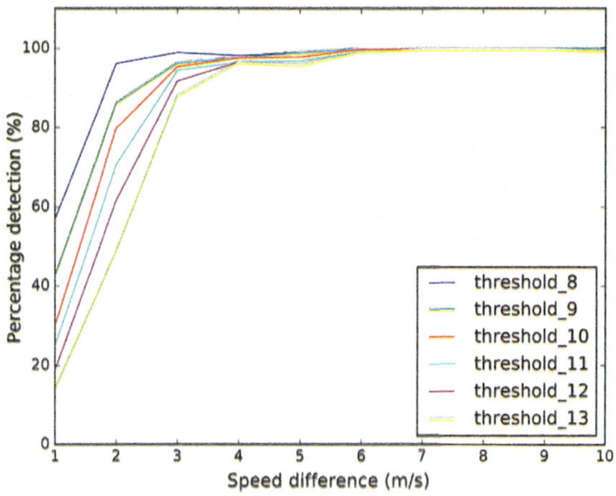

Fig. 6.16 Detection rate for East–West route

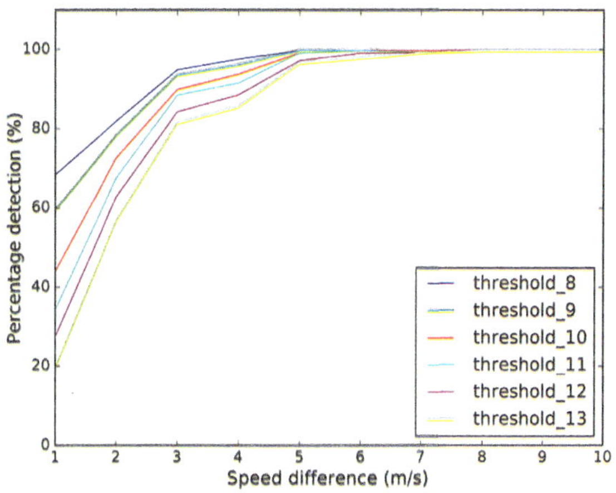

Fig. 6.17 Detection rate for North–South route

Six-mile Highway SC 29682 and 341 College Avenue, Clemson SC 29631). While traversing these different routes, we actively logged the difference in the calculated and measured Doppler shift values across the satellites. Given that the difference in calculated and measured Doppler shift values is not precisely zero even in the absence of spoofing attacks, we need a threshold value for the Doppler difference to determine when to issue an alarm for the detection of spoofing attacks. Of course, a larger threshold value will make it harder to detect spoofing attacks that result in the small Doppler shift discrepancies. Figures 6.16 and 6.17 show the rate of detection under different threshold values.

From Figs. 6.16 and 6.17, it can be seen that under all threshold values, the rate of detection increases as the difference increases between the spoofer's predicted velocity and the actual vehicle velocity. Moreover, for both the east–west route and the north–south route, beyond a speed difference of 6 m/s, the detection rate is almost always 100% for 500 runs based on different GPS chips and vehicle speeds. It is worth noting that for different routes, the relative direction of motions between the receiver and the satellites was different, which yielded slightly different detection rates. However, since in detection we used the average Doppler difference among the four satellites with the four most significant Doppler shift differences, the detection results should not be affected much by different routes. This explains similar statistical results for the two different routes and confirms the applicability of our approach irrespective of the route a vehicle takes.

6.6.2 *False Positives and Threshold Selection*

It is intuitive that a smaller threshold will lead to more false positive alarms, although it results in a higher detection rate. Given that it is unlikely for a vehicle to experience GPS spoofing attacks very often, sometimes it is reasonable to set the threshold to a value that gives the best detection rate under the constraint of zero false-alarm rate. We conducted over 500 tests to identify such a threshold value. Figure 6.18 shows the percentage of false positives for various thresholds when the receiver was not spoofed. It can be seen that if a threshold value of 13 is selected for the Doppler difference, then no false positive will exist. This means that in the absence of spoofing attacks, the calculated Doppler shift value will always be within 13 from the measured Doppler shift value reported by GPS receivers. (Recall that the Doppler difference is quantified using the average of four satellites.) Given that under threshold value 13, the detection rate is almost 100% when the speed difference is over 6 m/s, we recommend to enforce a purposed velocity variation of 6 m/s at times in traveling.

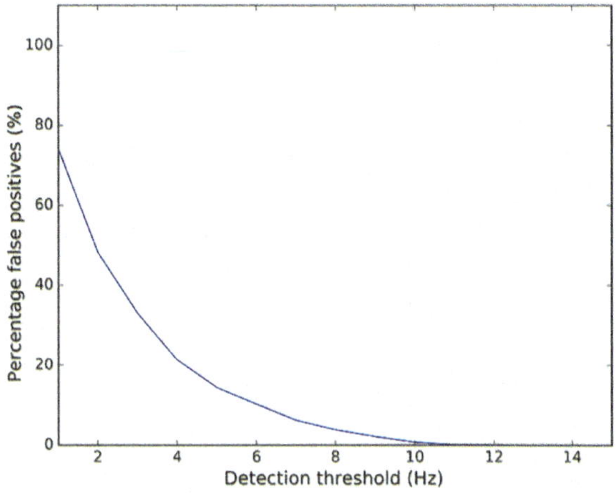

Fig. 6.18 False positive detection rate for different thresholds

6.6.3 Limitation of the Approach

Our approach counts on the inability of the spoofer to measure and incorporate the target velocity variation in real time in its generation of spoofing signals without any latency. In fact, during the latency in a spoofer's measuring target velocity/direction of traveling and adjusting fake signal frequency accordingly, purposed velocity changes we can enforce can lead to discrepancies between the actual Doppler shift value and spoofer's Doppler shift value, which will lead to the detection of spoofing attacks. As discussed earlier, the latency for spoofing is mainly caused by the delay in measuring the velocity and direction of traveling of a target vehicle (both are necessary to compute the Doppler effect), the delay in spoofer's generating and emitting fake signals at the right (time-varying) frequency, and the delay in commercial-grade GPS receivers' acquisition of a fix when message codes/contents are changed (to conduct spoofing). Since the first source of latency has been well studied in the literature (in, e.g., [29]), here we evaluate the latter two sources of delay in spoofing signal generation and acquisition. It is worth noting that although the open-source GPS simulation library GPS-SDR-SIM is used in almost all existing GPS spoofing studies, no results have been reported regarding its latency in generating GPS signals and spoofing commercial receivers. We conducted experiments to measure this latency and hence the potential limitations of our approach.

The GPS-SDR-SIM library requires ephemeris, time, and path coordinates to generate binary representations of GPS signals. The generated binary file is then passed on to a radio front end such as the HackRF or the USRP to broadcast the fake signal in the electromagnetic spectrum. An updated version of the GPS-SDR-SIM library includes a module that can predict and accommodate the Doppler shift frequency for supplied velocity measures. To circumvent our defending mechanism with enforced uncertain velocity variations, a spoofer must be able to update the spoofing signal frequency closely with the monitored motion variation. Here, assuming that the spoofer has direct access to vehicle velocity, we tested the latency of the GPS-SDR-SIM in generating spoofing signal files under changing receiver velocities. The GPS-SDR-SIM was implemented on an Intel Core i7 Quadcore 3.2 GHz processor with 8 GB of RAM. Its signal was then fed into the HackRF radio which was hard wired to spoof a Ublox NEO-M8T GPS module. We measured the time (latency) it took for the spoofer to spoof the Ublox in two cases.

6.6.3.1 Initial Takeover

We first tested the average time it took for the spoofer to take over the Ublox GPS receiver when the receiver was not locked onto any GPS signals. Figure 6.19 shows the times for the Ublox GPS receiver to have the first fix (lock onto a signal) in 50 trials.

The average time it took for the GPS receiver to lock onto the fake signal was 17.48 s and the minimum time was 14 s. This shows that the GPS-SDR-SIM has

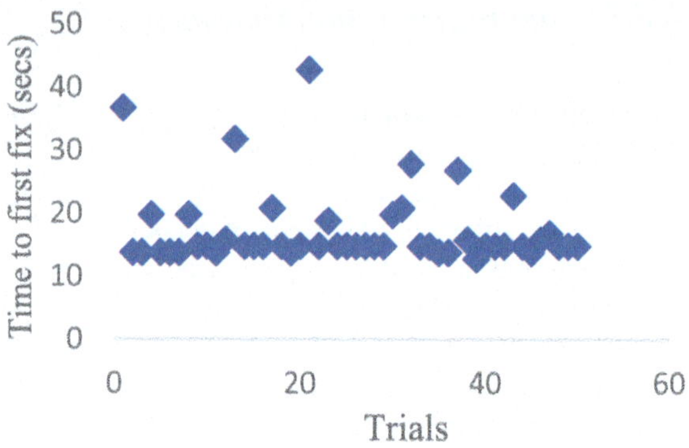

Fig. 6.19 Time to take over a GPS receiver that is not locked onto any GPS signal

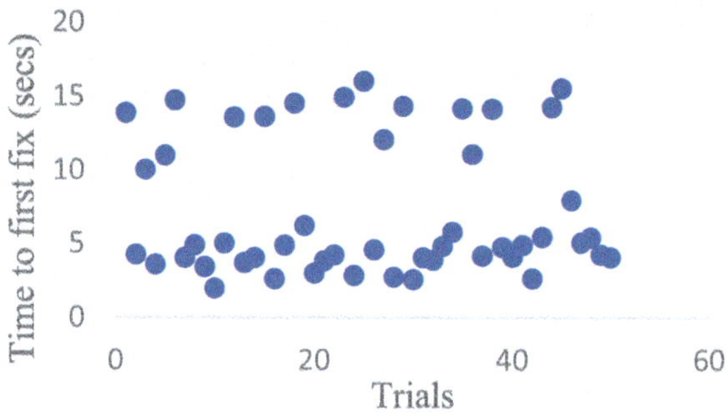

Fig. 6.20 Time to reacquire fix when the spoofer changes its frequency (the receiver was locked onto the spoofing signal before changes)

at least 14 s latency in successfully spoofing a receiver when the receiver was not locking onto any GPS signals. During such a latency, it is easy to enforce 6 m/s velocity variation which is sufficient to detect potential GPS spoofing attacks.

6.6.3.2 Continuous Takeover

Secondly, we considered a situation where the spoofer attempts to utilize the GPS-SDR-SIM to make the consecutive Doppler shift updates to keep the GPS receiver locked. The spoofer was still assumed to have access to the real-time velocity of the vehicle. We split a continuous spoofed data file with updated velocities and wrote an automated script to continually switch the spoofer through the updated parts to determine the minimum time it takes for the receiver to reacquire each of the updated broadcasts. On switching to the next file, the receiver loses fix momentarily and reacquires it relatively faster compared with the previous case. The receiver was able to lock faster because it was already fixed to the signal with valid satellites and only needed to recompute the position and time alone. Figure 6.20 shows the times it took for the receiver to reacquire the continuous signal in 50 trials.

It took for the GPS receiver an average of 7.3 s to lock onto the continuous spoofing signal and a minimum of 2.06 s. Also, it took for the GPS-SDR-SIM 0.3 s to generate a spoofing signal for the duration of 2.06 s. This means that the spoofer must predict the vehicle velocity for at least 2.36 s from the time it generates the fake signal to have a Doppler frequency consistent with the user's calculation. Given that in 2.36 s, it is easy to enforce a 6 m/s velocity variation to make a large enough inconsistency between the user-calculated and receiver-reported Doppler shift values (most commercially available vehicles can easily accelerate by $4.5\,\mathrm{m/s^2}$ [33]), our approach cannot be circumvented by GPS-SDR-SIM-based spoofers.

6.7 Conclusions

With GPS becoming increasingly important in numerous civilian applications, we are also facing increasingly severe security challenges due to the vulnerability of civilian GPS receivers. Despite plenty of proposed approaches to address the security of GPS receivers, we are still lacking cost-effective and easily deployable security solutions for commercial off-the-shelf GPS receivers from which raw GPS signals are inaccessible. We propose a GPS spoofing detection approach using the Doppler shift which is reported by the most existing commercial GPS receivers. To defeat sophisticated spoofers which might be able to accommodate the Doppler shift in generated fake signals, we propose to purposely enforce velocities variations that are unpredictable to the spoofer to yield the unpredictable Doppler shift. Experimental results show that even when the spoofer has access to the vehicle velocity in real time, it could still be detected by our detection approach due to the unavoidable latency in its generating spoofing signal and establishing fix on commercial GPS receivers. Our approach makes use of the velocity measurement from the OBD system, which is available on all modern cars manufactured after 1996, and is applicable to commercial GPS receivers that do not provide raw GPS signals. The avoidance of using specialized radios/ GPS receivers or receiver formations makes our approach promising for mass-market applications to protect commercial off-the-shelf GPS receivers, which is in a distinct difference from existing GPS spoofing detection approaches. Extensive field tests showed that the approach is insensitive to environmental changes such as weather conditions which might affect the Doppler shift measurements.

Acknowledgements The work was supported in part by the National Science Foundation under Grant 1738902.

References

1. Psiaki, M.L., O'Hanlon, B.W., Bhatti, J.A., Shepard, D.P., Humphreys, T.E.: GPS spoofing detection via dual-receiver correlation of military signals. IEEE Trans. Aerosp. Electron. Syst. **49**, 2250–2267 (2013)
2. Saarinen, J.: Students hijack luxury yacht with GPS spoofing. iTnews, 30 July 2013. https://www.itnews.com.au/news/students-hijack-luxury-yacht-withgps-spoofing-351659

3. Peterson, S.: Iran hijacked US drone, says Iranian engineer. Christian Science Monitor, 15 December 2011. https://www.csmonitor.com/World/Middle-East/2011/1215/Exclusive-Iran-hijacked-US-drone-says-Iranian-engineer
4. Hesch, J., Dimitrios, G.K., Bowman, S.L., Roumeliotis, S.I.: Camera-IMU-based localization: observability analysis and consistency improvement. Int. J. Robot. Res. **33**(1), 182–201 (2014)
5. Jansen, K., Schäfer, M., Moser, D., Lenders, V., Pöpper, C., Schmitt, J.: Crowd-GPS-Sec: Leveraging crowdsourcing to detect and localize GPS spoofing attacks. In: Proceedings IEEE Symposium on Security and Privacy, pp. 1018–1031 (2018)
6. J. A. Volpe National Transportation Systems Center: Vulnerability assessment of the transportation infrastructure relying on the global positioning system. Technical Report, US Dept. of Transportation (2008)
7. Ranganathan, A., Olafsdottir, H., Capkun, S.: SPREE: A spoofing resistant GPS receiver. In: Proceedings of the 22nd Annual International Conference on Mobile Computing and Networking, pp. 48–360 (2016)
8. Heng, L., Work, D.B., Gao, G.X.: GPS signal authentication from cooperative peers. IEEE Trans. Intell. Transp. Syst. **16**, 1794–1805 (2015)
9. Daneshmand, S., Jafarnia-Jahromi, A., Broumandan, A., Lachapelle, G.: A low-complexity GPS anti-spoofing method using a multi-antenna array. In: Proceedings ION GNSS12 Conference, pp. 1233–1243 (2012)
10. Scott, L.: Anti-spoofing & authenticated signal architectures for civil navigation systems. In: Proceedings of the 16th International Technical Meeting of the Satellite Division, pp. 1543–1552 (2003)
11. Wesson, K., Rothlisberger, M., Humphreys, T.: Practical cryptographic civil GPS signal authentication. J. Inst. Navig. **59**(3), 177–193 (2012)
12. Wullems, C.J.: A spoofing detection method for civilian L1 GPS and the E1-B Galileo safety of life service. IEEE Trans. Aerosp. Electron. Syst. **48**(4), 2849–2864 (2012)
13. Kuhn, M.: An asymmetric security mechanism for navigation signals. In: Proceedings of 6th International Workshop on Information Hiding, pp. 239–252 (2004)
14. Kerns, A., Shepard, D.P., Bhatti, J.A., Humphreys, T.E.: Unmanned aircraft capture and control via GPS spoofing. J. Field Robot. **31**(4), 617–636 (2014)
15. Sanders, C., Wang, Y.Q.: Localizing spoofing attacks on vehicular GPS using vehicle-to-vehicle communications. IEEE Trans. Veh. Technol. **69**(12), 15656–15667 (2020)
16. Lee, J.H., Kwon, K.C., An, D.S., Shim, D.: GPS spoofing detection using accelerometers and performance analysis with probability of detection. Int. J. Control Autom. Syst. **13**(4), 951–959 (2015)
17. Psiaki, M.L., Humphreys, T.E.: GNSS spoofing and detection. Proc. IEEE **104**(6), 1258–1270 (2016)
18. Tippenhauer, N.O., Papper, C., Rasmussen, K., Capkun, S.: On the requirements for successful GPS spoofing attacks. In: Proceedings of the 18th ACM Conference on Computer and Communications Security, pp. 75–86 (2011)
19. Papadimitratos, P., Jovanovic, A.: GNSS-based positioning: attacks and countermeasures. In: IEEE Military Communications Conference, pp. 1–7 (2008)
20. U-blox: U-blox M8 concurrent GNSS timing modules. https://www.u-blox.com/sites/default/files/NEO-LEA-M8T-FW3_DataSheet_%28UBX-15025193%29.pdf. [Accessed 20 11 2018]
21. SkyTraq: NS-RAW: carrier phase raw measurement output GPS receiver. NavSpark Store. http://navspark.mybigcommerce.com/ns-raw-carrier-phase-raw-measurement-output-gps-receiver/. [Accessed 21 11 2018]
22. Dr. Franz Fasching: RasPiGNSS Aldebaran
23. https://drfasching.com/products/raspignss/raspignss-aldebaran/. [Accessed 21 11 2018]
24. Swift Navigation: Piksi Multi GNSS Module. https://www.swiftnav.com/store/gnss-sensor-volume-orders/piksi-multi-gnss-module?utm_source=google&utm_medium=cpc&utm_term=&utm_content=general&utm_campaign=701F0000000gRQY&lp_content=general. [Accessed 21 11 2018]

25. Misra, P., Enge, P.: Global Positioning System: Signals Measurements and Performance. Gangna-Jamuna Press (2001)
26. Borre, K., Akos, D.M., Bertelsen, N., Rinder, P., Jensen, S. H.: A Software-Defined GPS and Galileo Receiver: A Single-Frequency Approach. Springer (2007)
27. Tsui, J.: Fundamentals of Global Positioning System Receivers. Wiley (2000)
28. Ebinuma, T., Kanemoto, S.: GPS-SDR-SIM. 24 Aug 2018. https://github.com/osqzss/gps-sdr-sim. [Accessed 20 11 2018]
29. Sim, S., Sock, J., Kwak, K.: Indirect correspondence-based robust extrinsic calibration of LiDAR and camera. Sensors **16**(6), 933 (2016)
30. Lehtinen, M., Happonen, A., Ikonen, J.: Accuracy and time to first fix using consumer-grade GPS receivers. In: Proceedings of 16th International Conference on Software, Telecommunications and Computer Networks, pp. 334–340 (2008)
31. Mastrigt, L.A., Wal A.J., Oonincx, P.J.: Exploiting the Doppler effect in GPS to monitor signal integrity and to detect spoofing. In: Proceedings of International Association of Institutes of Navigation World Congress (IAIN), pp. 1–8 (2015)
32. Harvey, A.C.: Forecasting, Structural Time Series Models and the Kalman Filter. Cambridge University Press, Cambridge (1990)
33. Barricella, M.: Acceleration of a car. The Physics Factbook (2001). https://hypertextbook.com/facts/2001/MeredithBarricella.shtml. (Accessed 23 11 2018)

Chapter 7
A System-of-Systems Approach to Strategic Cyber-Defense and Robust Switching Control Design for Cyber-Physical Wind Energy Systems

Juntao Chen and Quanyan Zhu

Abstract Due to the increasing adoption of smart and Internet of things (IoT) devices, Wind Energy System (WES) becomes more vulnerable to cyber- and physical-attacks. Therefore, designing a secure and resilient WES is critical. In this chapter, we first propose a System-of-Systems (SoS) framework for the cyber-physical WES. Specifically, on the one hand, we adopt a game-theoretic model to capture the interactions between the WES system defender and the adversary at the cyber-layer. The outcome of this cyber-defense game is reflected by control-aware Nash equilibria. On the other hand, we devise a cyber-aware robust and resilient switching controller based on a Markov jump linear system model for the physical WES. The performances of the WES cyber- and physical layers are interdependent due to their natural couplings. To design the integrated cyber-physical WES, we design an iterative algorithm to investigate the SoS equilibrium which considers the system security, robustness, and resilience in a holistic manner. Finally, we use case studies to corroborate the developed cross-layer design principles for the cyber-physical WES.

7.1 Introduction

Wind Energy System (WES) is a critical component of the smart grid to modernize the power grid. A significant amount of literature aims to control WES to achieve high reliability and efficient operation using various kinds of controllers [4]. One common feature of these works is that they have mainly focused on the optimal

J. Chen (✉)
Department of Computer and Information Science, Fordham University, New York, NY 10023, USA
e-mail: jchen504@fordham.edu

Q. Zhu
Department of Electrical and Computer Engineering, Tandon School of Engineering, New York University, Brooklyn, NY 11201, USA
e-mail: qz494@nyu.edu

H. Ishii and Q. Zhu (eds.), *Security and Resilience of Control Systems*,
Lecture Notes in Control and Information Sciences 489,
https://doi.org/10.1007/978-3-030-83236-0_7

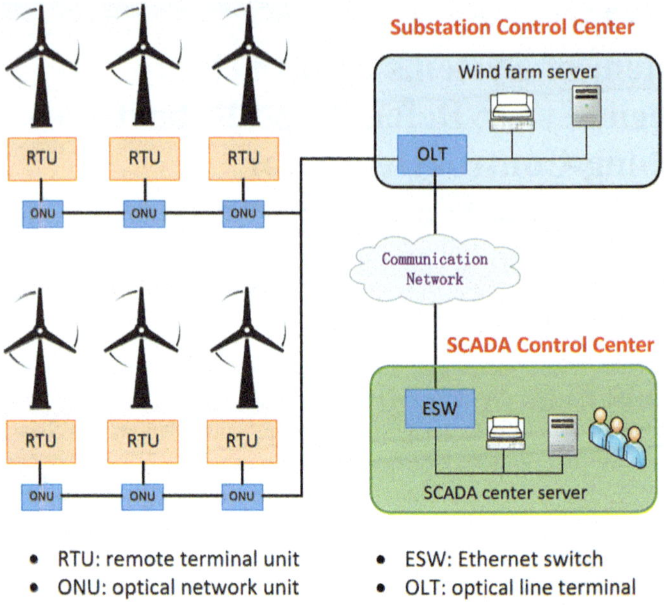

Fig. 7.1 A communication-based remote wind farm control system includes the SCADA system, communication networks, wind farm local server, and remote terminal units. This type of WES can be categorized to a networked control system or cyber-physical system

physical control design of a single wind turbine system. Due to the increasing demand for renewable energy, more and more large-scale wind farms are built [5]. One feature of wind farm operation is on its remote control through the support of sensing and communications [1]. Figure 7.1 depicts a framework of communication-based remote wind farm control system. Based on the data collected by the local sensors of WES, the Supervisory Control and Data Acquisition (SCADA) system first generates the commands and then sends them to the local WES server with the support of communication networks. Naturally, this type of WES can be regarded as a cyber-physical system.

The cybersecurity of WES becomes a critical concern as the integration of information technologies with the WES control systems. Based on the General Electric (GE) report [19], 96% wind farm has at least one machine with a vulnerable operation system, and hence the attackers can compromise the WES by attacking SCADA system and turbine controllers which leads to substantial financial losses, e.g., one day downtime at a 50 MW wind farm equals 29k$. Thus, strategic design is necessary to enhance the cybersecurity of WES. To achieve this goal, the WES control system operators need to design robust controllers to natural disturbances and develop secure defense mechanisms for WES cyber-system against adversaries. However, WES can hardly be perfectly secure and robust with the consideration of design costs. Hence, system resilience becomes critical for the WES to recover from the degraded state due to disturbances and adversaries.

Modern cyberattacks are elaborately designed and can be stealthy, e.g., Stuxnet attack [25]. The traditional firewall approach is not sufficient to defend against these cyberattacks [19]. Thus, cyber-defense in WES should shift from the focus on designing perfect security using firewall methods to best-effort strategic defense. To this end, our goal is to devise a secure, robust, and resilient cyber-physical WES from a cross-layer perspective. Specifically, we establish a noncooperative game-theoretic framework to capture the behaviors of system defender and cyber-attacker. We further establish a Markov Jump Linear System (MJLS) model for the WES where the state of MJLS represents the operation condition (e.g., normal or faulty) of the WES. Based on the MJLS model, a robust and resilient switching controller is designed to improve the WES performance. The WES resiliency is quantified by its recovery rates from the post-attack operation mode to the normal operation state. Note that cyber-defender's protection strategy governs the state transition of the WES which further influences the robust controller design. In addition, the outcomes of the established cyber-defense game have a direct impact on the WES operation efficiency. In sum, the two distinct cyber- and physical components of the WES are inherently *coupled*. Then, the designed strategies of cyber-players need to be control-aware, and in turn, the WES controller design at the physical layer should be cyber-aware. To overcome this challenge, we integrate the cyber- and physical layers into a *system-of-systems* framework to achieve a balanced design among the heterogeneous goals of WES system resilience, robustness, and security.

The main contributions of this chapter are presented as follows.

1. We develop an integrative system-of-systems framework for the holistic modeling of the cyber- and physical layers of the smart wind energy systems.
2. We design a robust and resilient stabilizing switching controller based on an MJLS model for the WES through a set of Linear Matrix Inequalities (LMIs). The controller considers the cyber-layer of WES and thus incorporates *cyber-aware* characteristic.
3. We provide the secure design principles of WES by analyzing a noncooperative game between the cyber-system defender and attacker. The *control-aware* Nash equilibrium solution of the established game is obtained by addressing Bilinear Programmings (BPs).
4. We formally define the *system-of-systems equilibrium* of the WES and design an algorithm to assess the cyber-physical WES performance under the developed cross-layer design methodology.

7.1.1 Related Work

Control of wind renewable energy systems has been studied extensively in literature with various types of intelligent controllers, and their resilient design plays a critical role in enhancing the smart grid efficiency [6, 29, 32, 35]. With the increasing integration of Information and Communication Technologies (ICTs) with smart grid, cyber-physical security becomes a critical issue in ensuring the efficient power system and critical infrastructure operations [28, 36, 37]. In [34], an integrated security framework based on multi-layer intrusion detection system has been designed for power grid automation. To protect the large-scale WES from cyberattacks, a security mechanism based on IEC 61400-25 communication model has been proposed in [26].

Cyber-Physical Systems (CPS) security and resilience have received a significant amount of attention in the last two decades. In [10], the authors have focused on the optimal design of multi-layer IoT-enabled CPS with heterogeneous security requirements. As cyber-layer is highly vulnerable to adversarial attacks, [13] has established a security as a service paradigm to ensure the trustworthy CPS operations. The authors in [20–23] have investigated the secure and resilient control of large-scale cyber-physical critical infrastructures, and designed efficient algorithms to compute the optimal solutions. Resilience of interdependent CPS has also been addressed comprehensively in [17]. Our work is related to [11], in which the authors have focused on the resilient control of WES using an MJLS model. However, security of cyber-system has not been addressed in [11].

Game-theoretic approach has been widely adopted to guide the secure system design and decentralized decision making, including computer networks [2, 27, 40], cyber-physical systems [9, 38], the Internet of things [15], and smart grids [12, 14]. Our cyber-physical cross-layer framework for secure and resilient control design is also related to [39, 41]. However, different from these literature, we focus on the specific WES cybersecurity problem and propose an integrative algorithm to achieve the optimal cyber-physical system-of-systems design.

CPS are inherently multi-layer in nature and are composed of heterogeneous decentralized components [8]. Hence, it requires system operators to secure the CPS from a *system-of-systems* perspective. In [30], the authors have proposed a Gestalt Nash equilibrium solution that integrates the security of cloud, communication, and physical layers in the modern IoT-enabled control applications. The authors in [16] have investigated the proactive security of multi-layer autonomous systems by proposing a meta-equilibrium solution. Furthermore, [24] has provided a comprehensive review of the joint-design of security and resilience of CPS from a cross-layer system-of-systems approach. Our current work is related to [31], where the authors have focused on the understanding of system-of-systems performance of coupled MJLS from a control-theoretic perspective.

7.1.2 Notation

The bold lower case $\mathbf{x}$ is a vector, and the bold upper case $\mathbf{X}$ represents a matrix. The notation $\mathbf{x} = [x_i]_{i \in \mathscr{S}}$ means that vector $\mathbf{x}$ is composed by the elements x_i where i belongs to the set $\mathscr{S}$. The parameter with superscripts $'$ and $*$ indicates the transpose

Table 7.1 Nomenclature

ϑ_{op}	Operating point of WES
ϑ_c	Operating condition of WES
ϑ_t	Inclusion of ϑ_{op} and ϑ_c, i.e., $\vartheta_t = [\vartheta_{op}, \vartheta_c]$
$\boldsymbol{\vartheta}_t$	Set of all admissible ϑ_t
$\mathscr{S}$	Set $\mathscr{S} := \{1, 2, ..., S\}$
$\tilde{\mathscr{S}}$	Set $\tilde{\mathscr{S}} := \{0, 1, ..., S-1\}$
γ_{ij}	Transition rate from state i to state j
a_i	Cyber-attacker's pure action
d_j	Cyber-defender's pure action
$\mathbf{g}$	Cyber-attacker's mixed strategy
$\mathbf{f}$	Cyber-defender's mixed strategy
P_{ij}	Power generation of WES under action pair $\{a_i, d_j\}$
$C_{a,i}$	Attack cost corresponding to action a_i
$C_{d,j}$	Defense cost corresponding to action d_j

and Nash equilibrium, respectively. For clarity, some critical notations adopted in this chapter are summarized in Table 7.1.

7.1.3 *Organization of the Chapter*

The rest of the chapter is organized as follows. Section 7.2 establishes a system-of-systems framework for the cyber-physical WES. Section 7.3 presents system modeling and cyberattack models. We design a cyber-aware robust and resilient switching controller in Sect. 7.4. We establish a secure cyber-system design framework and analyze a control-aware cyber-defense game in Sect. 7.5. Section 7.6 proposes an iterative algorithm to compute the solution. Section 7.7 presents case studies to corroborate the obtained results. Section 7.8 concludes the chapter.

7.2 System-of-Systems Framework

To capture the interdependencies between the cyber- and physical components of WES, it first requires establishing an integrated system framework.

As depicted in Fig. 7.2, the modern WES is naturally composed by the physical system and the cyber-system. At the physical layer, it contains the mechanical parts of WES and hence is related to the output of renewable energy. The cyber-layer plays an important role in the remote control of WES, and it is composed of the communication and sensor networks. On the one hand, the cyber-layer determines

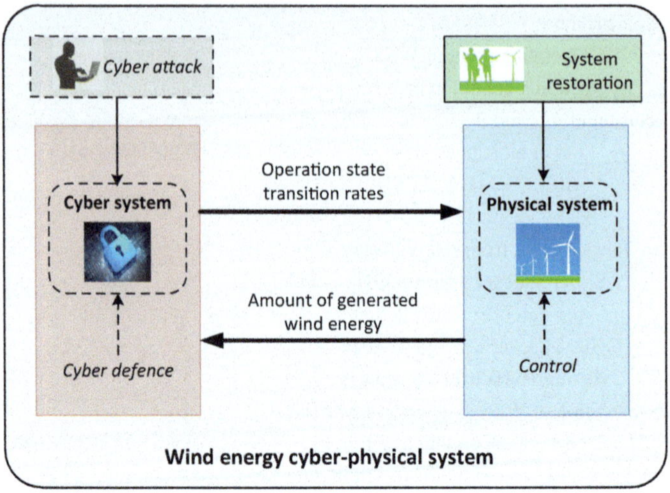

Fig. 7.2 A system-of-systems framework for modeling interdependencies in cyber-physical WES. The cyber-layer contains the communication and sensing components. The physical layer includes the mechanical and restoration parts

the *operating condition* of the WES. Specifically, the WES operates normally under no adversary, and the system is robust to disturbances with the designed controllers. In contrast, the WES operates in an inefficient post-attack mode after the cyber-attacker compromising the system. On the other hand, the cyber-layer is dependent on the physical layer performance as the *utilities* of the cyber-defender, and the adversary are related to the generated energy of the WES. Therefore, the cyber- and physical layers in WES are *interdependent*.

To this end, we propose a *system-of-systems* framework as shown in Fig. 7.2 to model the cyber-physical interdependencies. In addition to the adversarial attacks to the cyber-layer, the framework also incorporates a restoration component that captures the WES resilience. The couplings at different layers require us to devise efficient WES design from a cross-layer perspective. Therefore, in Sect. 7.6, we propose a *system-of-systems equilibrium* solution concept which guides a holistic design of secure, robust, and resilient cyber-physical WES.

7.3 WES Modeling and Cyberattack Models

First, we present the modeling of WES. Then, we introduce three classes of cyberattacks to the cyber-physical WES which can be integrated with the established WES model.

7.3.1 Modeling of WES

To model a WES, we need to consider its aerodynamic system, pitch system, electrical power system, drive train system, and the external wind speed. We briefly describe the WES model here, and more details are referred to [4].

7.3.1.1 Wind Speed

Wind speed $v(t)$ (m/s) is captured by the following two terms:

$$v(t) = v_m(t) + \hat{v}(t), \tag{7.1}$$

where $v_m(t)$ is the mean wind speed, and $\hat{v}(t)$ is the turbulence representing the high-frequency variable component of the wind. The turbulence part $\hat{v}(t)$ is often modeled by the Gaussian white noise.

7.3.1.2 Aerodynamic System

The aerodynamic power P_r (W) captured by the rotor is

$$P_r(t) = \frac{\pi \rho R^2}{2} C_p(\lambda, \beta) v(t)^3, \tag{7.2}$$

where ρ is the air density (kg/m^3); R is the radius of wind turbine rotor (m); and C_p is the power coefficient which depends on the pitch angle β and the tip speed ratio λ. Here, $\lambda = \omega_r R/v$, where ω_r is the rotor speed (rad/s). The rotor torque $T_r(t)$ (N·m) is given by

$$T_r(t) = \frac{P_r}{\omega_r} = \frac{\pi \rho R^3}{2} C_q(\lambda, \beta) v(t)^2, \tag{7.3}$$

where $C_q = \frac{C_p}{\lambda}$ is the torque coefficient.

7.3.1.3 Pitch System

A second-order dynamical system is used to model the hydraulic pitch actuator system, and

$$\ddot{\beta}(t) = -2\zeta \omega_n \dot{\beta}(t) - \omega_n^2 \beta(t) + \omega_n^2 \beta_d, \tag{7.4}$$

where ω_n is the natural frequency (rad/s), ζ is the damping ratio, and β_d is the commanded pitch angle. Additionally, the values of $\beta(t)$ and its slew rate $\dot{\beta}(t)$ both are bounded.

7.3.1.4 Drive Train System

The drive train includes a gear box, which gears up the rotation speed of rotor to a higher generator speed, and the purpose is to transfer torque from the rotor side to the generator side. Its dynamics are described as follows:

$$J_r\dot{\omega}_r = T_r - K_{dt}\theta_\Delta - (B_{dt} + B_r)\omega_r + \frac{B_{dt}}{N_g}\omega_g, \tag{7.5}$$

$$J_g\dot{\omega}_g = \frac{B_{dt}}{N_g}\omega_r + \frac{K_{dt}}{N_g}\theta_\Delta - \left(\frac{B_{dt}}{N_g^2} + B_g\right)\omega_g - T_g, \tag{7.6}$$

$$\dot{\theta}_\Delta = \omega_r - \frac{1}{N_g}\omega_g, \tag{7.7}$$

where J_r is the rotor inertia (kg·m^2); J_g is the generator inertia (kg·m^2); T_g is the generator torque (N·m); ω_g is the generator speed (rad/s); B_r is the rotor friction coefficient (N·m·s/rad); B_g is the generator friction coefficient (N·m·s/rad); K_{dt} is the torsion stiffness (N·m·s/rad); B_{dt} is the torsion damping coefficient (N·m·s/rad); θ_Δ is the torsion angle.

7.3.1.5 Electrical System

The generator and converter belong to the electrical part of the system, and their dynamics are much faster than the mechanical dynamics. For simplicity, it can be modeled as a first-order system

$$\dot{T}_g = -\frac{1}{\tau_g}T_g + \frac{1}{\tau_g}T_{gd}, \tag{7.8}$$

where τ_g is the time constant, and T_{gd} is the commanded generator torque (N·m).

7.3.2 WES System Representation

For the purpose of controller design, a linearized model of the system is convenient. The major nonlinear part in the system is the aerodynamic torque. We linearize (7.3) around the operating point $\vartheta_{op} := (\bar{v}, \bar{\beta}, \bar{\omega}_r)$, and arrive at

$$\hat{T}_r(\vartheta_{op}) = K_r(\vartheta_{op})\hat{\omega}_r + K_v(\vartheta_{op})\hat{v} + K_\beta(\vartheta_{op})\hat{\beta}, \tag{7.9}$$

where

$$K_r(\vartheta_{op}) = \left.\frac{\partial T_r}{\partial \omega_r}\right|_{\vartheta_{op}} = \frac{\rho\pi R^2 v^2}{2\omega_r}\left(R\frac{\partial C_p(\lambda,\beta)}{\partial\lambda} - C_p(\lambda,\beta)\frac{v}{\omega_r}\right),$$

$$K_v(\vartheta_{op}) = \left.\frac{\partial T_r}{\partial v}\right|_{\vartheta_{op}} = \frac{\rho\pi R^2 v}{2}\left(3C_p(\lambda,\beta)\frac{v}{\omega_r} - R\frac{\partial C_p(\lambda,\beta)}{\partial\lambda}\right),$$

$$K_\beta(\vartheta_{op}) = \left.\frac{\partial T_r}{\partial \beta}\right|_{\vartheta_{op}} = \frac{\rho\pi R^2 v^3}{2\omega_r}\frac{\partial C_p(\lambda,\beta)}{\partial\beta},$$

and $\hat{T}_r$, $\hat{v}$, $\hat{\beta}$ and $\hat{\omega}_r$ are the deviations from the operating point. Note that the operating point of the WES is time-varying due to the stochastic wind. The linearized WES model is given by

$$\begin{aligned}\dot{x}(t) &= A(\vartheta_t)x(t) + B(\vartheta_t)u(t) + E(\vartheta_t)\hat{v}(t),\\ y(t) &= C(\vartheta_t)x(t),\end{aligned} \tag{7.10}$$

where $A(\vartheta_t)$, $B(\vartheta_t)$, $E(\vartheta_t)$, and $C(\vartheta_t)$ are system matrices with appropriate dimensions; $x(t)$ is the state vector; $u(t)$ is the control input; $\hat{v}(t)$ is the exogenous perturbation; and $y(t)$ is the measured output. In addition, $\vartheta_t := [\vartheta_{op}, \vartheta_c] \in \boldsymbol{\vartheta}_t$ includes the operating point ϑ_{op} and operating condition ϑ_c of the wind turbine, where $\boldsymbol{\vartheta}_t$ is a set of all admissible ϑ_t. Note that the operating condition ϑ_c of WES is determined by the cyber-layer of the system-of-systems framework in Fig. 7.2. For example, under the normal operating condition, the state-space model of WES can be captured by

$$\begin{aligned}\dot{x}(t) &= \begin{bmatrix} -\frac{1}{\tau g} & 0 & 0 & 0 & 0 & 0\\ 0 & 0 & 1 & 0 & 0 & 0\\ 0 & -\omega_n^2 & -2\zeta\omega_n & 0 & 0 & 0\\ 0 & a_{42} & 0 & a_{44} & \frac{B_{dt}}{N_g J_r} & -\frac{K_{dt}}{J_r}\\ -\frac{1}{J_g} & 0 & 0 & \frac{B_{dt}}{N_g J_g} & a_{55} & \frac{K_{dt}}{N_g J_g}\\ 0 & 0 & 0 & 1 & -\frac{1}{N_g} & 0 \end{bmatrix} x(t)\\ &+ \begin{bmatrix} \frac{1}{\tau g} & 0\\ 0 & 0\\ 0 & \omega_n^2\\ 0 & 0\\ 0 & 0\\ 0 & 0 \end{bmatrix} u(t) + \begin{bmatrix} 0\\ 0\\ 0\\ \frac{K_v(\bar{V},\bar{\beta},\bar{\omega}_r)}{J_r}\\ 0\\ 0 \end{bmatrix} \hat{v}(t),\\ y(t) &= \begin{bmatrix} 1&0&0&0&0&0\\ 0&1&0&0&0&0\\ 0&0&0&1&0&0\\ 0&0&0&0&1&0 \end{bmatrix} x(t),\end{aligned} \tag{7.11}$$

where $x = [\hat{T}_g, \hat{\beta}, \dot{\hat{\beta}}, \hat{\omega}_r, \hat{\omega}_g, \hat{\theta}_\Delta]^T$, $u = [\hat{T}_{gd}, \hat{\beta}_d]^T$, $a_{42} = K_\beta(\vartheta_{op})/J_r$, $a_{44} = -(B_{dt}+B_r)/J_r + K_r(\vartheta_{op})/J_r$, $a_{55} = -(B_{dt}+N_g B_g)/(N_g^2 J_g)$ and $y = [\hat{T}_g, \hat{\beta}, \hat{\omega}_r, \hat{\omega}_g]^T$.

When the cyber-layer shown in Fig. 7.2 is not secure, the WES model (7.10) needs modification to incorporate the impacts of cyberattacks on the physical layer, which is detailed in the following section.

7.3.3 Adversarial Attacks Modeling

In the following, we present three classes of cyberattacks targeted at WES.

7.3.3.1 Sensor Attacks

One of the major types of malicious attacks in cyber-physical systems is the sensor attack [7, 33]. For instance, the adversary can compromise the sensing information in a strategic manner during the transmission through a man-in-the-middle attack. This type of attack can be captured using a sensor gain proportional modification model. In WES, we consider the following two sensor attacks: (i) pitch angle measurement sensor attack and (ii) generator speed measurement sensor attack. The sensor attack (i) yields modifications of the modeling of pitch system as well as the pitch angle measurement. Equivalently, the corresponding elements in matrices $B(\vartheta_t)$ and $C(\vartheta_t)$ are altered. To capture sensor attack (ii), we modify the matrices $A(\vartheta_t)$ and $C(\vartheta_t)$ in a similar way.

7.3.3.2 Actuator Attack

The second type of cyberattack targets the WES actuator. The adversary intrudes on the internal WES system and compromises the power converter controller. The mechanism of this actuator attack is similar to the Stuxnet [25], where the physical control system is corrupted. This actuator attack leads to an error on the generators' output torque, which is further reflected by the adjustment of $A(\vartheta_t)$ and $B(\vartheta_t)$.

7.3.3.3 Controller Attack

As shown in Fig. 7.1, WES is based on the remote control in which the SCADA system decides the control policies and then sends them to the local controller of WES via communications. The integration of ICTs makes the cyber-physical WES vulnerable to attacks [37]. For example, the attackers can launch an attack by compromising the control signal stealthily through invading to the SCADA system. This controller attack yields a modification of $B(\vartheta_t)$ in (7.10).

Remark The considered cyberattacks can be integrated with the established WES model (7.10) by modifying the associated system matrices. Also, multiple attacks may happen simultaneously, and their impacts on the system modeling should be considered jointly in the corresponding matrices. Note that under different cyberattacks, the physical WES operates under various conditions. This adversarial attack needs to be incorporated in the control design.

7.4 Robust and Resilient Cyber-Aware Switching Control Design

The operating condition of WES is related to the considered cyberattacks. In this section, we design a switching controller for WES based on its Markov jump linear system (MJLS) representation which aims to improve the WES performance.

7.4.1 MJLS Model of WES

The cyber-physical WES operates at various operating conditions captured by $\vartheta_c \in \boldsymbol{\vartheta}_c := \{\vartheta_n, \vartheta_{pa}, \vartheta_{gs}, \vartheta_{ac}, \vartheta_{ss}\}$, where ϑ_n, ϑ_{pa}, ϑ_{gs}, ϑ_{ac}, and ϑ_{ss} denote the normal, pitch angle sensor attack, speed sensor attack, actuator attack, and controller attack, respectively. Note that the dynamic operating point ϑ_{op} of WES is determined by the wind speed which admits a value in an appropriate subset of $\mathbb{R}^+$. One method for WES controller design is using gain-scheduling by choosing a set of operating points. In this way, the complexity of the WES controller depends on the cardinality of the selected operating point set. In this chapter, we specifically focus on the rated output power range of WES which leads to a fixed assumed ϑ_{op}. This focused scenario indicates that the variance of wind speed is limited, and the references $\bar{\beta}$ and $\bar{\omega}_r$ can be determined before the control design. Therefore, the parameters $K_r(\vartheta_{op})$, $K_v(\vartheta_{op})$, and $K_\beta(\vartheta_{op})$ in (7.9) can be seen as constants. However, the modeling can be extended to time-varying operating point scenarios depending on the number of chosen operating points during gain-scheduling. Note that though the reference of operating point can be seen as a prior, the WES operating condition is time-varying and depends on the real-time cyber-state, which introduces challenges to the control design.

The model presented in (7.10) can be regarded as a MJLS for WES under different operating conditions. We denote by $\{\vartheta_t, t \geq 0\}$ a right-continuous Markov chain on the complete probability space $(\Omega, \mathscr{F}, \{\mathscr{F}_t\}_{t\geq 0}, \mathbb{P})$, where Ω is the sample space; $\mathscr{F}$ is a σ-algebra of subsets of Ω; $\mathscr{F}_t$ is a filtration; and $\mathbb{P}$ is a probability measure on the measurable space $(\Omega, \mathscr{F})$. ϑ_t takes value in $\mathscr{S} = \{1, 2, ..., S\}$ with generator $\Gamma = \{\gamma_{ij}\}_{i,j\in\mathscr{S}}$ given by

$$\mathbb{P}\{\vartheta_{t+\Delta} = j \mid \vartheta_t = i\} = \begin{cases} \gamma_{ij}\Delta + o(\Delta), & \text{if } i \neq j, \\ 1 + \gamma_{ii}\Delta + o(\Delta), & \text{if } i = j, \end{cases} \tag{7.12}$$

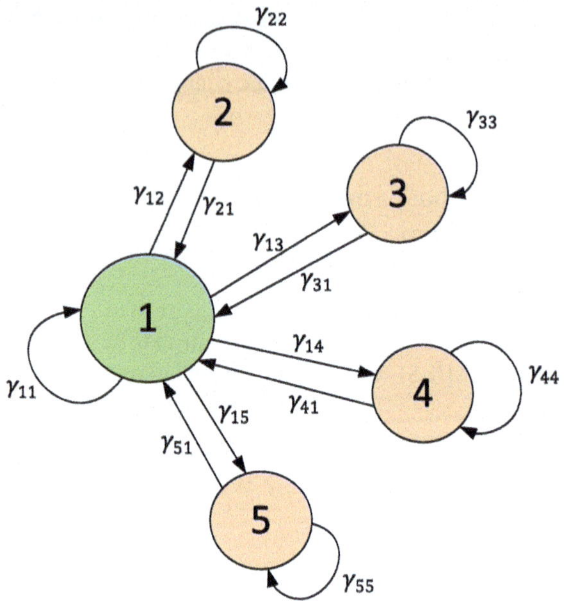

Fig. 7.3 State transition representation of the WES. State 1 is the normal operation, and states 2 to 5 represent the system operates under the pitch angle sensor attack (ϑ_{pa}), speed sensor attack (ϑ_{gs}), actuator attack (ϑ_{ac}), and controller attack (ϑ_{ss}), respectively. Also, $\{\gamma_{12}, \gamma_{13},...,\gamma_{15}\}$ and $\{\gamma_{21}, \gamma_{31},...,\gamma_{51}\}$ capture the security and resilience of the WES, respectively

where $\Delta > 0$; $\gamma_{ij} \geq 0$ is the transition rate from state i to state j when $i \neq j$, and $\gamma_{ii} = -\sum_{j=1,j\neq i}^{S} \gamma_{ij}$. For clarity, Fig. 7.3 depicts the state transition of the WES. Specifically, the states from 1 to 5 representing the WES operates under the normal, pitch angle sensor attack, speed sensor attack, actuator attack, and controller attack scenarios, respectively. For the considered model, there are $S = 5$ operating conditions in total. The transition rates γ_{ij}, $i = 1, \forall j \neq 1 \in \mathscr{S}$, correspond to the attacker's behavior that leads to a failure of WES.

7.4.2 *Cyber-Aware Robust Switching Controller Design*

Our next goal is to design a robust switching controller for the WES based on the established MJLS model. When $\vartheta_t = i \in \mathscr{S}$, we denote $A(\vartheta_t) := A_i$ for convenience, and similar for $B(\vartheta_t)$, $C(\vartheta_t)$ and $E(\vartheta_t)$. The designed dynamic output feedback controller takes the following form:

$$\begin{aligned} \dot{x}_F(t) &= A_F(\vartheta_t)x_F(t) + B_F(\vartheta_t)y(t), \\ u(t) &= C_F(\vartheta_t)x_F(t), \end{aligned} \tag{7.13}$$

where $A_F(i) = A_{Fi}$, $B_F(i) = B_{Fi}$, and $C_F(i) = C_{Fi}$ are system matrices, $\forall i \in \mathscr{S}$; and $x_F(t)$ represents the state of the designed controller. Furthermore, the closed-loop system needs to have a level κ of H_∞ noise attenuation, i.e.,

$$\int_0^{\infty} \mathbb{E}(|z(t)|^2)dt < \kappa^2 \int_0^{\infty} \mathbb{E}(|\hat{v}(t)|^2)dt, \tag{7.14}$$

where κ is a prescribed constant, $\mathbb{E}(\cdot)$ denotes the expectation operator; $|\cdot|$ is the standard Euclidean norm; $\hat{v}(t)$ is the exogenous perturbation in (7.10); $z(t)$ is defined as $z(t) := Q(\vartheta_t)x(t) + R(\vartheta_t)u(t)$, and $Q(\cdot) \succeq 0$ and $R(\cdot) \succeq 0$. For $\vartheta_t = i \in \mathscr{S}$, we denote $Q(\vartheta_t) = Q_i$ and $R(\vartheta_t) = R_i$. Then, for $\forall i \in \mathscr{S}$, the dynamic controller (7.13) admits the following forms:

$$\begin{aligned} A_{Fi} &= (Y_i^{-1} - X_i)^{-1} M_i Y_i^{-1}, \\ B_{Fi} &= (Y_i^{-1} - X_i)^{-1} L_i, \\ C_{Fi} &= F_i Y_i^{-1}, \end{aligned} \tag{7.15}$$

where

$$M_i = -A_i' - X_i A_i Y_i - X_i B_i F_i - L_i C_i Y_i - Q_i'(Q_i Y_i + R_i F_i) - \kappa^{-2} X_i E_i E_i' - \sum_{j=1}^{S} \gamma_{ij} Y_j^{-1} Y_i.$$

The matrices X_i, Y_i, L_i, and F_i are the solutions to the following linear matrix inequalities (LMIs)

$$\begin{bmatrix} T_1 & Y_i'Q_i' + F_i'R_i' & \mathscr{R}_i(Y) \\ Q_i Y_i + R_i F_i & -I & 0 \\ \mathscr{R}_i'(Y) & 0 & \mathscr{D}_i(Y) \end{bmatrix} \prec 0, \quad \begin{bmatrix} T_2 & X_i E_i \\ E_i' X_i & -\kappa^2 I \end{bmatrix} \prec 0, \quad \begin{bmatrix} Y_i & I \\ I & X_i \end{bmatrix} \succ 0, \tag{7.16}$$

where

$$\begin{aligned} T_1 &:= A_i Y_i + Y_i A_i' + B_i F_i + F_i' B_i' + \gamma_{ii} Y_i + \kappa^{-2} E_i E_i', \\ T_2 &:= A_i' X_i + X_i A_i + L_i C_i + C_i' L_i' + Q_i' Q_i + \sum_{j=1}^{S} \gamma_{ij} X_j, \\ \mathscr{R}_i(Y) &:= [\sqrt{\gamma_{1i}} Y_i, \ldots, \sqrt{\gamma_{(i-1)i}} Y_i, \sqrt{\gamma_{(i+1)i}} Y_i, \ldots, \sqrt{\gamma_{Si}} Y_i], \\ \mathscr{D}_i(Y) &:= -\mathrm{diag}(Y_1, \ldots, Y_{i-1}, Y_{i+1}, \ldots, Y_S). \end{aligned}$$

Note that the feedback controller (7.13) depends on the system matrices as well as the measurement $y(t)$. The specific values of A_{Fi}, B_{Fi}, and C_{Fi} in (7.13) are obtained by solving LMIs in (7.16).

Under the designed switching feedback control, the WES switches between the designed robust controllers in (7.13) agilely based on the measured ϑ_t. An illustration

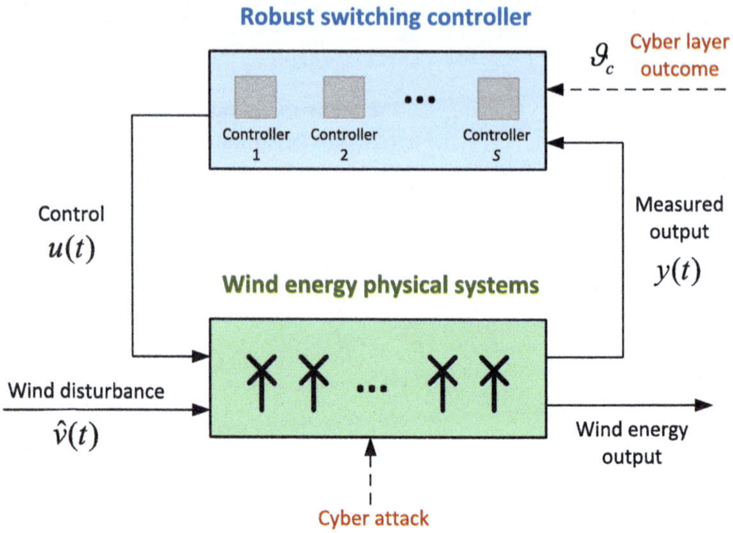

Fig. 7.4 The structure of the robust switching controller for WES. Based on the outcome of the cyber-layer which determines the operating condition ϑ_c, the WES can switch between different designed robust controllers swiftly

of the designed robust switching controller for WES is shown in Fig. 7.4. The real-time control depends on the cyber-layer outcome and the measured output from the WES. Based on the state ϑ_t, the controller admits the corresponding one to mitigate the adversarial impact on the energy system.

Remark The MJLS (7.10) coupled with the dynamic controller (7.13) with parameters (7.15) is mean-square stable [18], and this ensures the stability of the WES under adversary environment.

7.4.3 *WES Resilience*

In the MJLS model, the transition rates $\gamma_{ji}, i = 1,\ j \neq i \in \mathscr{S}$, represent the resilient ability of WES. The WES exhibiting a faster restoration from the failure state to the normal mode is more resilient to cyberattacks, and a larger γ_{ji} reflects this fact. Note that removing the cyberattacks is costly for the WES operator. Therefore, the operator needs to take into account the trade-off between the system recovery and the secure WES design costs. Recall that (7.15) is related to the system restoration transition rates, and thus the designed switching controller captures the resilience of the WES. For clarity, the transition rates $\gamma_{ji},\ i = 1$, are considered as fixed in the current work. The extension to incorporate unfixed restoration cost in which $\gamma_{ji},\ i = 1$ are decision variables is nontrivial and requires another layer of optimization.

7.5 Noncooperative Game-Theoretic Control-Aware Cyber-System Design

The cyber-layer of WES has a direct influence on its physical components as the cyber-layer security governs the physical system state transition rates. As perfect cybersecurity is costly prohibitive, the system operator needs to allocate his defensive resources optimally. To this end, it requires the operator to understand the strategic behavior of the cyber-attacker and thus correspondingly develop the defensive mechanism. To achieve this goal, we establish a noncooperative game-theoretic control-aware framework to capture the strategic interactions between the defender and attacker at the cyber-layer of WES. This framework facilitates the prediction of security risks of the cyber-system under strategic attacks.

7.5.1 *Control-Aware Cyber-Defense Game*

We denote by this cyber-game $\mathbf{G}_c$. The first step is to identify the action sets for both the attacker and the defender. Specifically, the modeled cyberattacks in Sect. 7.3.3 are sequentially denoted as $a_1,\ a_2,\ a_3$, and a_4, respectively. The SCADA system can monitor the wind farm operation and detect cyber-intrusions. According to the attack types, the defensive strategies can be categorized into sensor, actuator, and controller defense. Similarly, we denote the defender's action with respect to the corresponding attack by $d_1,\ d_2,\ d_3$, and d_4, respectively. In addition, both attacker and defender can be inactive which are denoted as a_0 and d_0. Therefore, the attacker's action a belongs to the set $\mathscr{A} := \{a_0, a_1, a_2, a_3, a_4\}$, and the defender's action d is chosen from $\mathscr{D} := \{d_0, d_1, d_2, d_3, d_4\}$. The mixed strategies of attacker and defender are $\mathbf{g} = [g_i]_{i=0}^{S-1} \in \mathscr{G}$ and $\mathbf{f} = [f_i]_{i=0}^{S-1} \in \mathscr{F}$, respectively, where g_i and f_i are the probabilities of choosing $a_i \in \mathscr{A}$ and $d_i \in \mathscr{D}$, respectively. The admissible strategies of two players are denoted as follows:

$$\begin{aligned} \mathscr{G} &:= \left\{ \mathbf{g} \in [0,1]^S : \sum_{i=0}^{S-1} g_i = 1 \right\}, \\ \mathscr{F} &:= \left\{ \mathbf{f} \in [0,1]^S : \sum_{j=0}^{S-1} f_j = 1 \right\}. \end{aligned} \tag{7.17}$$

The next step is to define the payoff functions of both players in the game $\mathbf{G}_c$. Under a pure strategy profile $\{a_i, d_j\}$, the WES's output power in a fixed time horizon is $P_{ij}, i, j \in \tilde{\mathscr{S}} := \{0, ..., S-1\}$. The costs of actions a_i and d_j for the attacker and defender are $C_{a,i}$ and $C_{d,j}$, respectively. Specifically, we have the null-strategy costs: $C_{a,0} = 0$ and $C_{d,0} = 0$. With strategy pair $\{a_i, d_j\}$, the defender's utility is

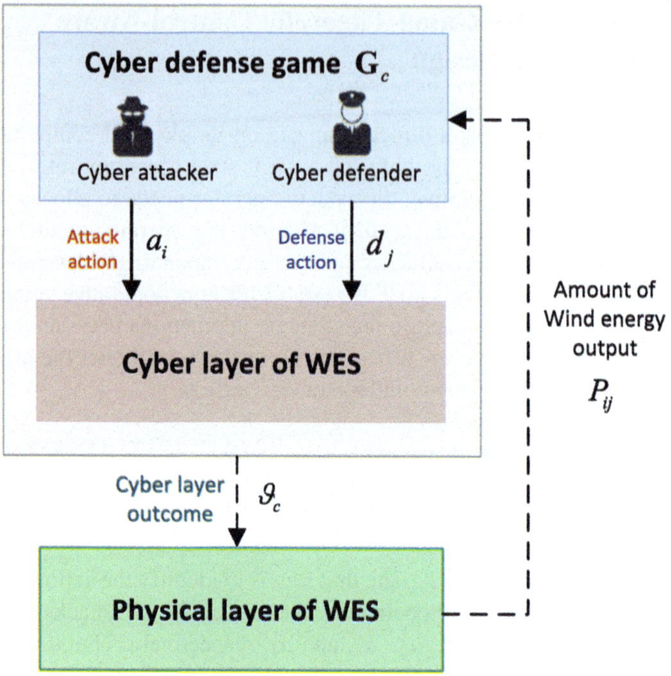

Fig. 7.5 A noncooperative game-theoretic framework that models the cybersecurity of WES. Both the attacker and the defender make strategic decisions to optimize their objectives, and their strategies are dependent on the wind energy output. Due to the inner coupling, the performance of the physical system is influenced by the security game outcome on the cyber-layer

$$U_{ij} = \alpha P_{ij} - C_{d,j}, \tag{7.18}$$

and the attacker's cost is

$$C_{ij} = \alpha P_{ij} + C_{a,i}, \tag{7.19}$$

for $i, j \in \tilde{\mathscr{S}}$, where α is a weighting constant. Note that the defender is a maximizer and the attacker is a minimizer in the established nonzero-sum cyber-game. The strategy outcome of the game determines the operation state transition rates $\gamma_{ij}, i = 1, \forall j \neq i \in \mathscr{S}$, of the WES in (7.12) which affects the performance of the wind energy physical system through (7.16). Figure 7.5 illustrates the game-theoretic framework that models the WES cybersecurity, and also shows the explicit interdependencies between the cyber-layer and physical energy system.

For a given cyber-strategy pair $\{\mathbf{g}, \mathbf{f}\}$, the transition rate γ_{1i} takes the form $\gamma_{1i} = \sum_{j \neq i-1, j \in \tilde{\mathscr{S}}} \eta f_j g_{i-1}$, for $i \neq 1 \in \mathscr{S}$, where η is a relatively small positive weighting constant transforming the probability to rate measure and thus γ_{1i} models realistic cyberattack frequencies. Further, $\gamma_{11} = -\sum_{i \neq 1 \in \mathscr{S}} \gamma_{1i}$. Note that γ_{1i} captures that fact that the attack action a_{i-1} is successfully only when the defender's action is not d_{i-1}. The expected attack and defense costs of players are given by $C_a = \sum_{i \in \tilde{\mathscr{S}}} C_{a,i} g_i$, and $C_d = \sum_{j \in \tilde{\mathscr{S}}} C_{d,j} f_j$, respectively. The defender and attacker determine strategic decisions to trade-off between the cyber-layer security and the incurred cost.

7.5.2 *Control-Aware Nash Equilibrium*

A natural solution concept to characterize the strategies of two players with competing goals is *Nash Equilibrium* (NE) composed by $\{\mathbf{g}^*, \mathbf{f}^*\}$. Denote $\mathbf{U} = [U_{ij}]_{\forall i, j \in \tilde{\mathscr{S}}}$ and $\mathbf{C} = [C_{ij}]_{\forall i, j \in \tilde{\mathscr{S}}}$. Then, the definition of NE in mixed strategies of the cyber-game is as follows.

Definition 7.1 (*Control-Aware NE of Cyber-Game* $\mathbf{G}_c$) A strategy pair $\{\mathbf{g}^* \in \mathscr{G}, \mathbf{f}^* \in \mathscr{F}\}$ constitutes a noncooperative NE of the cyber-defense game $\mathbf{G}_c$ in mixed strategies if

$$\mathbf{f}'\mathbf{U}\mathbf{g}^* \le \mathbf{f}^{*\prime}\mathbf{U}\mathbf{g}^*, \text{ and } \mathbf{g}^{*\prime}\mathbf{C}\mathbf{f}^* \le \mathbf{g}'\mathbf{C}\mathbf{f}^*, \tag{7.20}$$

for $\forall \mathbf{f} \in \mathscr{F}$, and $\forall \mathbf{g} \in \mathscr{G}$.

Remark: Under the NE, the expected payoffs are equal to $\bar{U} := \mathbf{f}^{*\prime}\mathbf{U}\mathbf{g}^*$ and $\bar{C} := \mathbf{g}^{*\prime}\mathbf{C}\mathbf{f}^*$ for the defender and attacker, respectively. The existence of mixed strategy NE of the cyber-game $\mathbf{G}_c$ is ensured by Nash's Theorem [3]. In addition, the mixed strategy NE can be obtained through solving a Bilinear Programming (BP).

7.6 Interdependent WES Analysis and Design

Designing cyber- and physical layers of cyber-physical systems is generally separate. However, as shown in Sects. 7.4 and 7.5, the design of an efficient WES requires a holistic method by integrating the cyber-system design and the physical controller design together. Therefore, the switching control and the defensive strategy of cyber-physical WES are interdependent.

In this section, we first define a system-of-systems equilibrium solution concept for the integrated system and then propose an iterative algorithm to obtain this equilibrium which guides the holistic design of cyber-physical WES.

7.6.1 *System-of-Systems Equilibrium*

The NE security strategies of game $\mathbf{G}_c$ in Sect. 7.5 is computed by BP, and the LMI facilitates the switching controller design in Sect. 7.4. The resulting controller of LMI impacts the average WES output power P_{ij}, $\forall i, j \in \tilde{\mathscr{S}}$, over $t \in [0, T]$. Then,

LMI's outcome, generated power, provides an input for the BP in forming the game $\mathbf{G}_c$ at the cyber-layer. Note that the NE of $\mathbf{G}_c$ determines the state transition rates γ_{ij}, $\forall i, j \in \mathscr{S}$, and consequently impacts the LMI in (7.16). In sum, we can predict the outcome of the coupled cyber- and physical layers in WES by finding a *consistent solution* which satisfies BP and LMI at the same time. To achieve this goal, we propose the following concept of *system-of-systems equilibrium* to assess the design of the interdependent cyber-physical WES.

Definition 7.2 (*System-of-Systems Equilibrium*) The system-of-systems equilibrium in the interdependent cyber-physical WES is a point that the cyber-strategy and the switching controller satisfy the following conditions.

(i) The cyber-strategy pair $\{\bar{\mathbf{g}}, \bar{\mathbf{f}}\}$ is an NE of the cyber-defense game $\mathbf{G}_c$ that corresponds to a wind power output profile $\{P_{ij}\}$, for $i, j \in \tilde{\mathscr{S}}$;
(ii) Based on the state transition rates γ_{ij}, $i, j \in \mathscr{S}$, resulting from the cyber-strategy $\{\bar{\mathbf{g}}, \bar{\mathbf{f}}\}$, the designed switching controller (7.13) of WES yields a consistent wind power output profile $\{P_{ij}\}$ with that in (i), for $i, j \in \tilde{\mathscr{S}}$.

The system-of-systems equilibrium requires us to design the cyber- and physical layers in WES holistically.

7.6.2 *Iterative Algorithm*

To obtain the system-of-systems equilibrium, we develop an iterative algorithm based on solving the interleaving LMIs and BPs. For convenience, a flow diagram of the algorithm is shown in Fig. 7.6. We summarize the iterative algorithm in the following. First, we initialize a cyber-strategy from (7.17), and determine the robust switching control of WES by solving LMIs (7.16). We update the game $\mathbf{G}_c$ as the utilities of attacker and defender are changed, and then recompute the NE strategy which further facilitates the next round of update. Through a number of alternating updates of these two parts, the robust switching controller at the system-of-systems equilibrium is obtained, and the strategies of the game converge to $\bar{\mathbf{g}}$ and $\bar{\mathbf{f}}$ for the attacker and defender, respectively. We will show the effectiveness of the proposed design strategy using case studies in Sect. 7.7.

7.7 Case Studies

We use case studies to corroborate the obtained results in this section. Specifically, we construct a WES with system parameters specified in [29]. The average wind speed is 15 m/s, and its disturbance is an independent and identical Gaussian distribution with variance 0.3. For illustrative clarity, we investigate two operation conditions without loss of generality, i.e., under normal operation and pitch angle sensor attack.

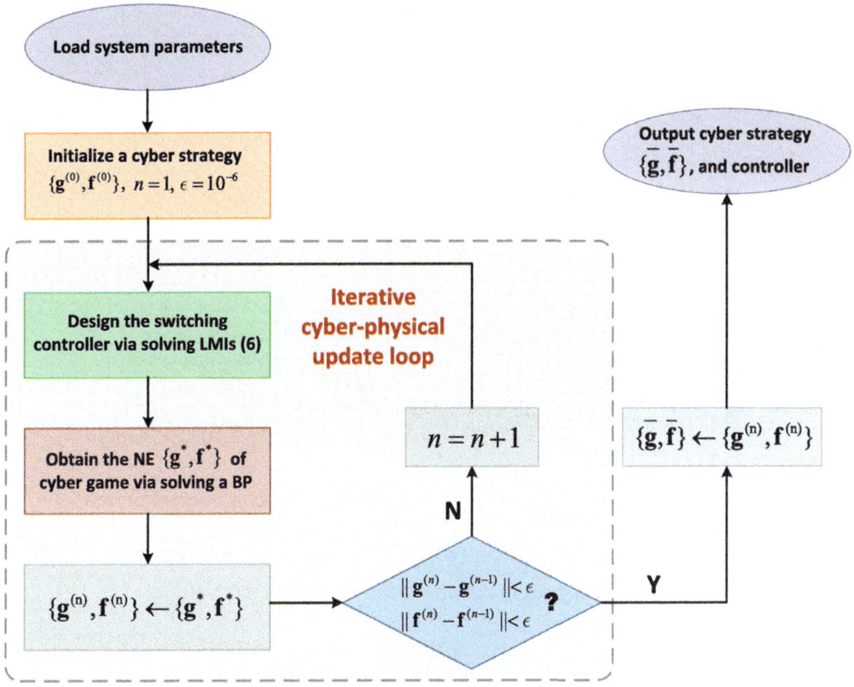

Fig. 7.6 A flow diagram of the iterative algorithm to obtain the system-of-systems performance. The iterative loop contains the switching controller design for the physical layer and defense strategy update for the cyber-layer

Note that other attacks can be incorporated similarly. In the focused scenarios, the attacker corrupts the pitch angle sensor, and injects a constant gain 0.8 to the sensor value β. Thus, the control-aware cyber-defense game is two-dimensional. Also, we assume that the detection of the change of operating condition is agile, e.g., through analyzing the received time series sensor information at the terminal control unit. And the magnitude of sensor corruption by the adversary can be inferred through statistical difference between normal and compromised data using inference techniques. In the following studies, we will show the robustness, security and resilience of WES in different scenarios, and provide WES control guidelines under adversarial environment.

7.7.1 *Cyber-Aware Robust Switching Control*

We first corroborate the designed cyber-aware robust switching controller, and the robustness level is chosen as $\kappa = 10$ in (7.14). The results are depicted in Fig. 7.7. Figure 7.7a shows the sampled two-state Markov chain under the transition rate matrix

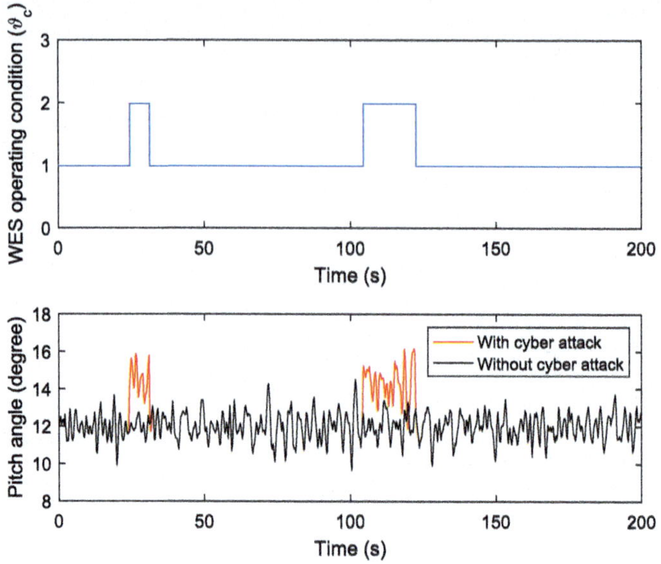

(a) WES operating condition and measured pitch angle.

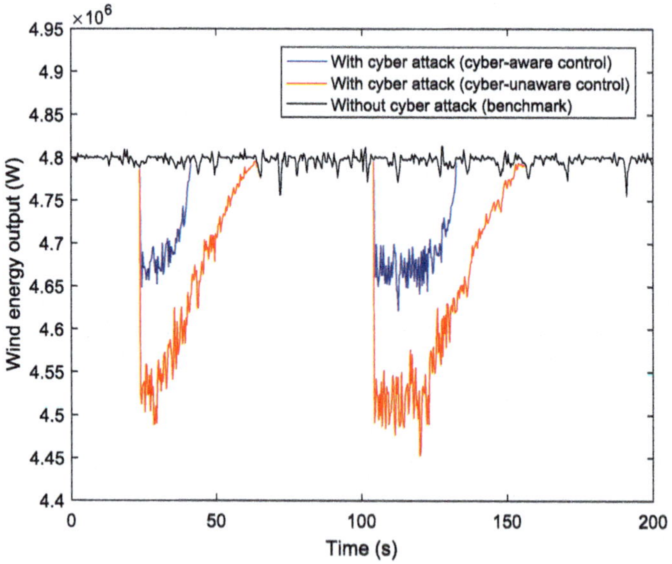

(b) WES output power.

Fig. 7.7 **a** Shows the operating condition of a WES in a certain period of time including the normal and pitch angle sensor attack states. **b** Depicts the corresponding WES output power under the designed cyber-aware switching control together with cyber-unaware control for comparison

$$\Gamma = \begin{bmatrix} -0.003 & 0.003 \\ 0.1 & -0.1 \end{bmatrix},$$

which governs the WES operating condition. Note that the values of transition rates vary across different energy systems, and they can be determined specifically by the historical data of the attack frequencies in targeted WES. Figure 7.7b shows the WES output power using the designed control. For comparison, the WES performance under cyber-unaware control is also illustrated in which the controller only considers WES normal operation. The results show that the cyber-aware switching control outperforms the one without awareness significantly. When the cyberattack happens, the switching control is more efficient since it incorporates this security consideration in the design beforehand. Based on the statistic [19], the monetary loss for a 50 MW wind farm using the designed cyber-aware control can be reduced to around 10k$ per month compared with the cyber-unaware counterpart, and this shows the advantages of the proposed methodology.

7.7.2 *Impact of Security and Resilience*

We next investigate the impact of cybersecurity on the WES performance. Specifically, we compare two cases in which the energy systems (WES 1 and WES 2) are with transition rates

$$\Gamma_1 = \begin{bmatrix} -0.001 & 0.001 \\ 0.1 & -0.1 \end{bmatrix} \text{ and } \Gamma_2 = \begin{bmatrix} -0.005 & 0.005 \\ 0.1 & -0.1 \end{bmatrix},$$

respectively. Note that the cyber-layer of WES 1 is more secure than that of WES 2 due to a smaller γ_{12}, while the resilience of two systems is the same (same γ_{21}). The corresponding WES output power is shown in Fig. 7.8. We can see that WES 1 outperforms WES 2 in the output power, and thus a securer cyber-system enhances the WES efficiency. This study suggests that, if the system operator has more budget, then he can invest more on the cyber-layer of WES to reduce the risks of compromise of the cyber-system.

To study the system resilience, we compare two scenarios in which the energy systems' transition rates are as follows:

$$\Gamma_1 = \begin{bmatrix} -0.001 & 0.001 \\ 0.1 & -0.1 \end{bmatrix} \text{ and } \Gamma_2 = \begin{bmatrix} -0.001 & 0.001 \\ 0.04 & -0.04 \end{bmatrix}.$$

These two systems are of the same level of security (same γ_{12}), while the former one is more resilient to attacks. The obtained result is shown in Fig. 7.9. We can see that the first WES can recover from the cyberattack more quickly than the latter one, and hence a better system performance is achieved.

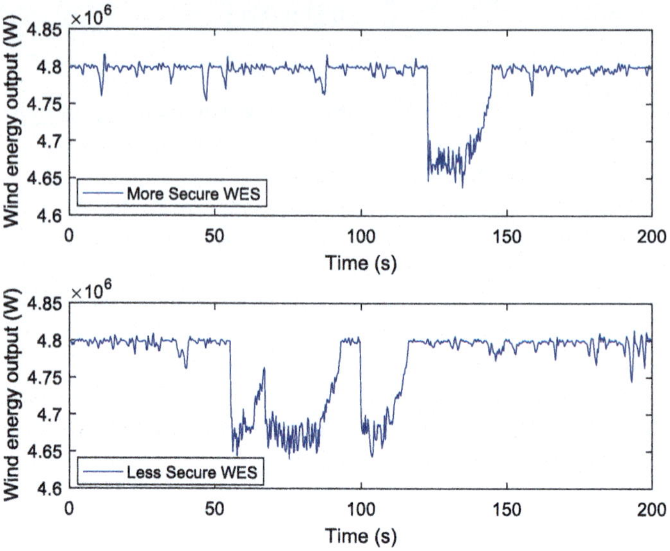

Fig. 7.8 WES performance under different cybersecurity levels. WES 1 outperforms WES 2, i.e., with more output, due to a higher security level

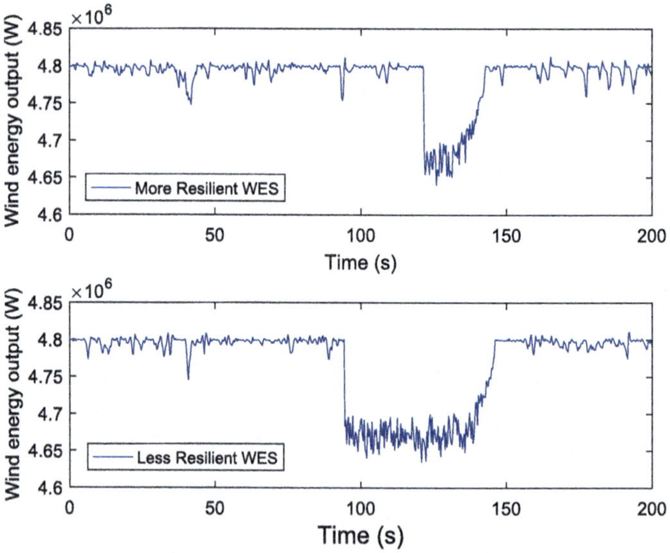

Fig. 7.9 WES performance under different resilience levels. The WES with a better resilience ability (former one) can yield more output power

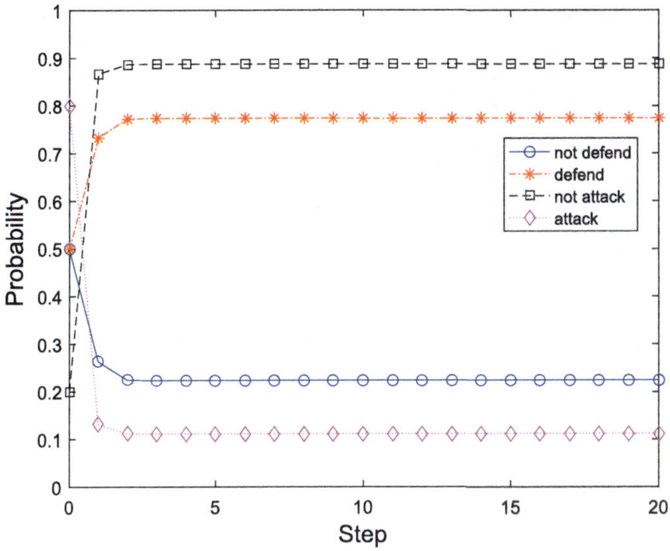

Fig. 7.10 Illustration of the iterative update algorithm. The algorithm converges to mixed cyber-strategies $\bar{\mathbf{g}} = (0.895, 0.105)$ and $\bar{\mathbf{f}} = (0.224, 0.776)$ for the attacker and defender, respectively

In conclusion, increasing either γ_{21} or γ_{12} can improve WES efficiency. The system designer needs to allocate the constrained cybersecurity resources and physical restoration resources strategically to obtain the desired outcome. This fact also demonstrates the trade-off decision-making between system security and resilience.

7.7.3 Integrated Cyber-Physical WES Design

In this subsection, we corroborate the effectiveness of the proposed iterative algorithm that yields a system-of-systems equilibrium design. Specifically, the cost parameters in the cyber-game are assumed to be $C_{a,1} = 10$k\$ and $C_{d,1} = 5$k\$, and the weighting factor admits $\alpha = 10^{-2}$\$/W. We then compute the average generated renewable energy P_{ij} under the designed control for given strategy pair $\{a_i, d_j\}$, $i, j \in \{0, 1\}$. Based on the iterative algorithm, Fig. 7.10 depicts the updated strategies of the defender and attack in the control-aware cyber-defense game. The cyber-game strategies are iteratively evolved by considering the physical WES performance. The algorithm successfully results in a system-of-systems equilibrium after several rounds of updates at which the players' strategies are $\bar{\mathbf{f}} = (0.224, 0.776)$ and $\bar{\mathbf{g}} = (0.895, 0.105)$, respectively. This equilibrium strategy guides the strategic cyber-defense of WES operator by considering the incurred physical system performance.

7.8 Conclusion

We have proposed a system-of-systems framework for the holistic design of cyber-physical WES that jointly takes the system robustness, security, and resilience into account. For the cyber-layer operator, we have provided a strategic defense mechanism using a noncooperative cyber-game. For the physical WES operation, we have designed a switching controller which has been corroborated to be robust and resilient. The system-of-systems equilibrium of the WES has been computed using an iterative mechanism by solving BPs and LMIs alternatively. For future work, we can explicitly quantify the values of robustness, security, and resilience of the designed system and use them for the optimal design of cyber-physical WES under adversaries.

Acknowledgements This research is partially supported by award 2015-ST-061-CIRC01, U. S. Department of Homeland Security (DHS), awards ECCS- 1550000, ECCS-1847056, CNS-1544782, and SES-1541164 from National Science of Foundation (NSF).

References

1. Ahmed, M.A., Kim, Y.C.: Communication network architectures for smart-wind power farms. Energies **7**(6), 3900–3921 (2014)
2. Alpcan, T., Başar, T.: Network Security: A Decision and Game-Theoretic Approach. Cambridge University Press (2010)
3. Başar, T., Olsder, G.J.: Dynamic Noncooperative Game Theory, Classics in Applied Mathematics, vol. 23. SIAM (1999)
4. Bianchi, F.D., De Battista, H., Mantz, R.J.: Wind Turbine Control Systems: Principles, Modelling and Gain Scheduling Design. Springer, Berlin (2006)
5. Bolinger, M., Wiser, R.: Wind power price trends in the united states: struggling to remain competitive in the face of strong growth. Energy Policy **37**(3), 1061–1071 (2009)
6. Burkart, R., Margellos, K., Lygeros, J.: Nonlinear control of wind turbines: an approach based on switched linear systems and feedback linearization. In: Proceedings of IEEE Conference on Decision and Control and European Control Conference, pp. 5485–5490 (2011)
7. Cardenas, A.A., Amin, S., Sastry, S.: Secure control: Towards survivable cyber-physical systems. In: Proceedings of IEEE Conference on Distributed Computing Systems, pp. 495–500 (2008)
8. Chen, J., Touati, C., Zhu, Q.: Heterogeneous multi-layer adversarial network design for the IoT-enabled infrastructures. In: Proceedings of IEEE Global Communications Conference, pp. 1–6 (2017)
9. Chen, J., Touati, C., Zhu, Q.: A dynamic game approach to strategic design of secure and resilient infrastructure network. IEEE Trans. Inf. Forensics Sec. **15**, 462–474 (2019)
10. Chen, J., Touati, C., Zhu, Q.: Optimal secure two-layer IoT network design. IEEE Trans. Control Netw. Syst. **7**(1), 398–409 (2020)
11. Chen, J., Zhou, L., Zhu, Q.: Resilient control design for wind turbines using Markov jump linear system model with Lévy noise. In: Proceedings of IEEE International Conference on Smart Grid Communications (SmartGridComm), pp. 828–833 (2015)
12. Chen, J., Zhu, Q.: A game-theoretic framework for resilient and distributed generation control of renewable energies in microgrids. IEEE Trans. Smart Grid **8**(1), 285–295 (2016)

13. Chen, J., Zhu, Q.: Security as a service for cloud-enabled internet of controlled things under advanced persistent threats: a contract design approach. IEEE Trans. Inf. Forensics Sec. **12**(11), 2736–2750 (2017)
14. Chen, J., Zhu, Q.: A Stackelberg game approach for two-level distributed energy management in smart grids. IEEE Trans. Smart Grid **9**(6), 6554–6565 (2017)
15. Chen, J., Zhu, Q.: Interdependent strategic security risk management with bounded rationality in the Internet of Things. IEEE Trans. Inf. Forensics Sec. **14**(11), 2958–2971 (2019)
16. Chen, J., Zhu, Q.: Control of multi-layer mobile autonomous systems in adversarial environments: a games-in-games approach. IEEE Trans. Control Netw. Syst. **7**(3), 1056–1068 (2020)
17. Chen, J., Zhu, Q.: A Game-and Decision-Theoretic Approach to Resilient Interdependent Network Analysis and Design. Springer (2020)
18. de Farias, D.P., Geromel, J.C., do Val, J.B., Costa, O.L.V.: Output feedback control of Markov jump linear systems in continuous-time. IEEE Trans. Autom. Control **45**(5), 944–949 (2000)
19. GE: GE Cybersecurity for Wind. [Online]. https://www.ge.com/content/dam/gepower-renewables/global/en_US/downloads/brochures/ge-wind-cybersecurity.pdf (Accessed April 2020)
20. Huang, L., Chen, J., Zhu, Q.: A factored MDP approach to optimal mechanism design for resilient large-scale interdependent critical infrastructures. In: Proceedings of Workshop on Modeling and Simulation of Cyber-Physical Energy Systems (MSCPES), pp. 1–6 (2017)
21. Huang, L., Chen, J., Zhu, Q.: A large-scale Markov game approach to dynamic protection of interdependent infrastructure networks. In: Proceedings of International Conference on Decision and Game Theory for Security, pp. 357–376. Springer (2017)
22. Huang, L., Chen, J., Zhu, Q.: Distributed and optimal resilient planning of large-scale interdependent critical infrastructures. In: Proceeding of Winter Simulation Conference (WSC), pp. 1096–1107 (2018)
23. Huang, L., Chen, J., Zhu, Q.: Factored Markov game theory for secure interdependent infrastructure networks. In: Proceedings of Game Theory for Security and Risk Management, pp. 99–126. Springer (2018)
24. Huang, Y., Chen, J., Huang, L., Zhu, Q.: Dynamic games for secure and resilient control system design. Nat. Sci. Rev. **7**(7), 1125–1141 (2020)
25. Langner, R.: Stuxnet: dissecting a cyberwarfare weapon. IEEE Sec. Privacy **9**(3), 49–51 (2011)
26. Liu, N., Zhang, J., Liu, W.: A security mechanism of web services-based communication for wind power plants. IEEE Trans. Power Deliv. **23**(4), 1930–1938 (2008)
27. Manshaei, M.H., Zhu, Q., Alpcan, T., Başar, T., Hubaux, J.P.: Game theory meets network security and privacy. ACM Comput. Surv. (CSUR) **45**(3), 25 (2013)
28. Mo, Y., Kim, T.H.J., Brancik, K., Dickinson, D., Lee, H., Perrig, A., Sinopoli, B.: Cyber-physical security of a smart grid infrastructure. Proc. IEEE **100**(1), 195–209 (2012)
29. Odgaard, P.F., Stoustrup, J., Kinnaert, M.: Fault-tolerant control of wind turbines: a benchmark model. IEEE Trans. Control Syst. Technol. **21**(4), 1168–1182 (2013)
30. Pawlick, J., Chen, J., Zhu, Q.: iSTRICT: an interdependent strategic trust mechanism for the cloud-enabled Internet of controlled things. IEEE Trans. Inf. Forensics Sec. **14**(6), 1654–1669 (2018)
31. Peng, G., Chen, J., Zhu, Q.: Distributed stabilization of two interdependent Markov jump linear systems with partial information. IEEE Control Syst. Lett. **5**(2), 713–718 (2021)
32. Shao, C., Wang, X., Shahidehpour, M., Wang, X., Wang, B.: Security-constrained unit commitment with flexible uncertainty set for variable wind power. IEEE Trans. Sustain. Energy **8**(3), 1237–1246 (2017)
33. Teixeira, A., Pérez, D., Sandberg, H., Johansson, K.H.: Attack models and scenarios for networked control systems. In: Proceedings of the 1st International Conference on High Confidence Networked Systems, pp. 55–64 (2012)
34. Wei, D., Lu, Y., Jafari, M., Skare, P., Rohde, K.: An integrated security system of protecting smart grid against cyber attacks. In: Proc. Innovative Smart Grid Technologies (ISGT), pp. 1–7 (2010)

35. Wu, B., Lang, Y., Zargari, N., Kouro, S.: Power Conversion and Control of Wind Energy Systems. Wiley (2011)
36. Wu, G., Sun, J., Chen, J.: A survey on the security of cyber-physical systems. Control Theory Technol. **14**(1), 2–10 (2016)
37. Yan, Y., Qian, Y., Sharif, H., Tipper, D.: A survey on cyber security for smart grid communications. IEEE Commun. Surv. Tutor. **14**(4), 998–1010 (2012)
38. Yuan, Y., Yuan, H., Guo, L., Yang, H., Sun, S.: Resilient control of networked control system under DoS attacks: a unified game approach. IEEE Trans. Ind. Inf. **12**(5), 1786–1794 (2016)
39. Yuan, Y., Yuan, H., Ho, D.W.C., Guo, L.: Resilient control of wireless networked control system under denial-of-service attacks: A cross-layer design approach. IEEE Trans. Cybernet. **50**(1), 48–60 (2020)
40. Zhang, R., Zhu, Q., Hayel, Y.: A bi-level game approach to attack-aware cyber insurance of computer networks. IEEE J. Sel. Areas Commun. **35**(3), 779–794 (2017)
41. Zhu, Q., Başar, T.: Game-theoretic methods for robustness, security, and resilience of cyber-physical control systems: games-in-games principle for optimal cross-layer resilient control systems. IEEE Control Syst. Mag. **35**(1), 46–65 (2015)

Chapter 8
Machine Learning-Based and Physics-Based Attack Resilient Wide-Area Monitoring, Protection and Control Systems

Gelli Ravikumar and Manimaran Govindarasu

Abstract The electric power grid is a complex critical infrastructure that forms the lifeline of modern society. Securing networked Cyber-Physical Systems (CPSs), such as the power grid critical infrastructure, has emerged as paramount importance to national security and economic security. Adversaries exploit the vulnerabilities in both cyber- and physical-domains and deploy stealthy multistage cyberattacks to disrupt reliable operation of the smart grid. Recent cyberattacks (in 2015 and 2016) that targeted the Ukrainian power grid and targeted cyberattack campaign during 2017–19 demonstrate the need for techniques to detect and mitigate such attacks. These incidents demonstrated growing threats and vulnerabilities within the smart grid, where critical control centers present a major attack target and whose compromise could result in major impacts on grid operation, including blackouts. This chapter presents cyberattack surface analysis for CPS wide-area control system and proposes an attack-resilient framework for supervisory control and data acquisition, energy management system and Wide-Area Monitoring, Protection and Control (WAMPAC) applications of bulk power grid. We discuss Machine Learning (ML) based anomaly detection and mitigation for a WAMPAC application, namely Wide-Area Damping Control (WADC) as case study. We present the architecture of the Hardware-in-the-Loop (HIL) CPS security power grid testbed and discuss the experimental results for the case study.

8.1 Introduction

The modern grid leverages cyber-infrastructure to enable wide-area situational awareness and secure system operation afforded by Wide-Area Monitoring, Protection and Control (WAMPAC) capabilities. Power grid control centers worldwide

G. Ravikumar (✉) · M. Govindarasu
Iowa State University, Ames, IA, USA
e-mail: gelli@iastate.edu

M. Govindarasu
e-mail: gmani@iastate.edu

H. Ishii and Q. Zhu (eds.), *Security and Resilience of Control Systems*,
Lecture Notes in Control and Information Sciences 489,
https://doi.org/10.1007/978-3-030-83236-0_8

equipped with various wide-area controls and sensor technologies across the grid to achieve the WAMPAC while improving system operation, resiliency, and efficiency. The cyber-infrastructure, including Wide-Area Communication Networks (WANs) across the geographically distributed sensors and control center applications, plays a crucial role in providing low latency and high bandwidth requirements. High latency and low bandwidth significantly affect the performance of wide-area time-critical applications such as state estimation, real-time contingency analysis, remedial action schemes, automatic generation control, and Wide-Area Damping Control (WADC).

On the other hand, cyberattacks such as a Denial of Service (DoS), Distributed DoS (DDoS), unauthorized switching of breakers, replay, and data integrity on the compromised WANs impose maloperation of the WAMPAC applications. Moreover, securing the grid against stealthy cyberattacks is challenging due to the legacy nature of the infrastructure and the dynamic nature of the threat landscape, and the ever-growing sophistication of the adversaries. Recent cyberattacks (in 2015 and 2016) that targeted the Ukrainian power grid demonstrate the need for techniques to detect and mitigate such attacks [7, 15]. Furthermore, during 2017–19, a targeted cyberattack campaign was detected within the US energy infrastructure [6]. These incidents demonstrated pervasive vulnerabilities within modern power systems, where critical control centers present a major attack target and whose compromise can directly cause a major blackout. As utilities become increasingly interconnected to the Independent System Operator (ISO), Balancing Authorities (BA), and other utilities, these complex and highly interconnected networks introduce growing cybersecurity challenges.

The rest of the chapter is organized as follows. Section 8.2 discusses attack surface analysis for the CPS control loop. Section 8.3 proposes an attack-resilient framework for SCADA and WAMPAC systems. Section 8.4 presents the hardware-in-the-loop CPS security testbed. Section 8.5 discusses a case study on attack-resilient using Machine Learning, followed by conclusion in Sect. 8.6.

8.2 Attack Surface Analysis for CPS Control Loop

8.2.1 Overview of the CPS Architecture

Figure 8.1 shows a Cyber-Physical System (CPS) control loop for the modern power systems. The actuators and sensors form CPS interfaces between the power grid physical resources and control center applications. Actuators include Power System Stabilizers (PSSs) at generator power plants, controllers at Flexible AC Transmission Systems (FACTS), and controllers at High Voltage DC (HVDC) systems. Sensors include Phasor Measurement Units (PMUs) to measure voltage phasors via potential transformers and current phasors via Current Transformers (CTs). Besides the actuators and sensors, control center system operators may issue command signals

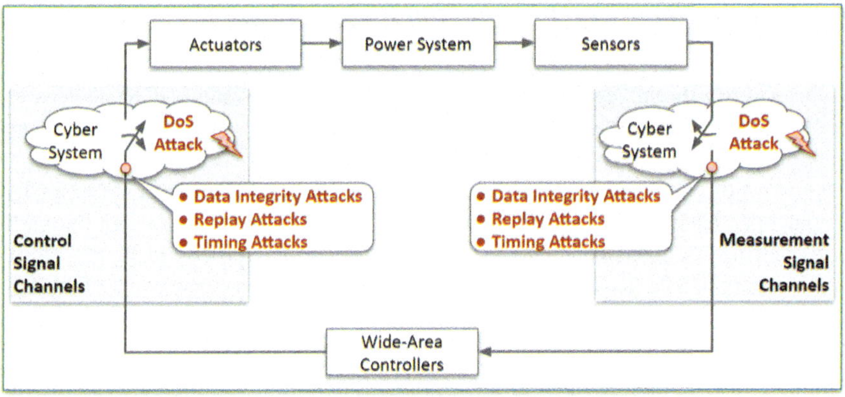

Fig. 8.1 CPS control loop architecture for wide-area smart grid applications

to relays or Intelligent Electronic Devices (IEDs) for toggling Circuit Breaker (CB) statuses and modulate power grid topologies.

As the actuators, sensors, and relays/CBs provide monitoring and wide-area controls via wide-area communication networks, and adversaries may compromise the communication networks to impose maloperations on their desired functionality. Adversaries may launch cyberattacks on data-at-rest, which is at the field devices via malware and data-in-transit. The data on communication networks via DoS, DDoS Data integrity, replay, and timing attacks. Cybersecurity for data-in-transit is a challenging task as there are more vulnerabilities, various attack paths, and a lack of situational awareness on the wide-area communication infrastructure. The Security Information and Event Management (SIEM) tools are essential for CPS-wide situational awareness. WAMPAC systems are critical for the situational awareness and operational stability of the bulk power system [33]. WAMPAC essentially leverages the PMUs and existing SCADA measurements to get a real-time snapshot of power system operations and execute real-time protection and control applications including State Estimation (SE), Real-Time Contingency Analysis (RTCA), Remedial Action Schemes (RAS), Automatic Generation Control (AGC), and Wide-Area Damping Control (WADC) [5, 21, 33, 34]. While increased communication of the PMUs and SCADA devices is essential to enable wide-area smart grid functionalities, devising defense-in-depth and cybersecurity for the communication protocols such as DNP3, Modbus, and IEEE C37.118 against cyberattack vectors is crucial for cyber-secure and attack-resilient SCADA and WAMPAC applications. We also note that co-simulation [27] and Hardware-in-the-Loop (HIL) CPS security testbed [17, 22, 23] are essential for the analysis of inter-dependent dynamics involved in power and cyber-systems for various CPS smart grid applications subject to various cyberattacks and grid impact characteristics.

8.2.2 CPS Security Challenges and Requirements

Cybersecurity threats against various Industrial Control System (ICS) devices and utility-wide CPS assets have been realized for decades [16]. In the postmortem analysis of the terrorist cyberattacks on September 11, 2001, utilities recognized the critical need for cybersecurity for critical infrastructures. Advanced threats, such as Stuxnet [9], have shown the ability to produce attacks that encompass traditional IT and control system environments to target physical system processes. Compromised and insecure ICS devices may lead to catastrophic disruptions, disclosure of sensitive information, and frauds [13]. Adversaries exploit the vulnerabilities in utility-wide cyber-systems to launch cyberattacks, which result in a cyber-threat to the utility's cyber-infrastructure. Potential cyber-threats to both ICS and SCADA systems [3] directly affect cyber-systems and indirectly affect power systems. A primary concern has been the possibility of High Impact Low Frequency (HILF) [4] cyberattacks on ICS, SCADA, and WAMS communication control systems and the resulting impacts on the overall observability, controllability, and stability. Such attacks in turn affect the resiliency of power systems and can lead to cascading outages or blackouts.

Extensive research has been carried out to unfold the cybersecurity posture and system-wide vulnerabilities for the power systems and critical infrastructures [2, 10, 12, 19]. The advent of WAMPAC and smart grid technologies and the essential requirement of system-wide substantial information exchanges raise the significance of cybersecurity. Attack resiliency is one of the primary requirements for the next-generation cyber-secure modern power systems, as reported by the U.S. Department of Energy (DOE) [8, 11, 14, 18]. Cybersecurity guidelines for the smart grid, including the advent technologies of Distributed Energy Resources (DER) and Wide Area

Table 8.1 Cybersecurity challenges and solutions for control center applications

Control center networks properties	Cybersecurity challenges	Proposed solutions
Networks are used to support a large number of integrated power applications, including control (AGC) functions	Unclear how malicious messages could (i) immediately impact grid operations, or (ii) have propagative impacts to interrelated functions	Develop a causal and predictive method to determine potential system impacts of malicious activity on power applications
Networks must transmit and process over 100,000 data points, collected every 4–60 s	Difficult to perform manual analysis of potential anomalies and understand their grid impacts	Require automated techniques to compute statistical models over high-dimensional data
Grid applications depend on many external variables including weather, energy costs, and demand	Machine learning techniques will result in a large number of anomalies from normal behavior	Require methods utilizing human feedback and reinforcement learning to reduce false positives and adapt to system changes

Table 8.2 Sample wide-area power grid applications and security

Application	Description
Wide-area Damping Control (WADC)	Demonstrate the feasibility of stealthy cyberattacks that manipulate WADC operation of the WAMPAC application, impact on inter-area oscillation modes, machine learning-based anomaly detection, and model-based mitigation for damping inter-area oscillations [21]
Automatic Generation Control (AGC)	Demonstrate the feasibility of AGC false data injection attacks that manipulate grid frequency and model-based defenses to ensure acceptable ranges [31]
SCED	Demonstrate controller memory attacks that manipulate transmission line ratings and resulting economic dispatch [30]

Measurement Systems (WAMS), have been reported by the NIST (National Institute for Standards and Technology) as NISTIR-7628 standard [35]. Table 8.1 identifies specific functional properties of control center networks, resulting in cybersecurity challenges, and the potential solutions to address them through attack detection, mitigation, and information sharing approaches (Table 8.2).

8.3 Attack-Resilient SCADA and WAMPAC

8.3.1 Proposed Attack-Resilient Framework for SCADA and WAMPAC

The proposed attack-resilient framework for SCADA and WAMPAC control center applications is shown in Fig. 8.2. We define the attack-resiliency is a function of detection, mitigation, prevention, and recovery of cyberattacks on the power grids. Moreover, the successful cyberattacks can affect the observability, controllability, and stability of power systems. It results in the suboptimal operation of power systems and may lead to blackouts. In the proposed framework, we introduced the CPS Resilience layer between the control center applications and the traditional Front End Processor (FEP), configuring grid data monitoring and control data points, and solicits SCADA data and PMU devices, and dispatches data to the relays/IEDs and actuators. The CPS Resilience layer consists of attack or anomaly detection and attack or anomaly mitigation algorithms and correlation and statistical analytics on cyber- and grid data streams. We proposed two designs for the CPS anomaly detection and mitigation system (CPS-ADMS): (1) Generic-ADMS and (2) Application-Specific

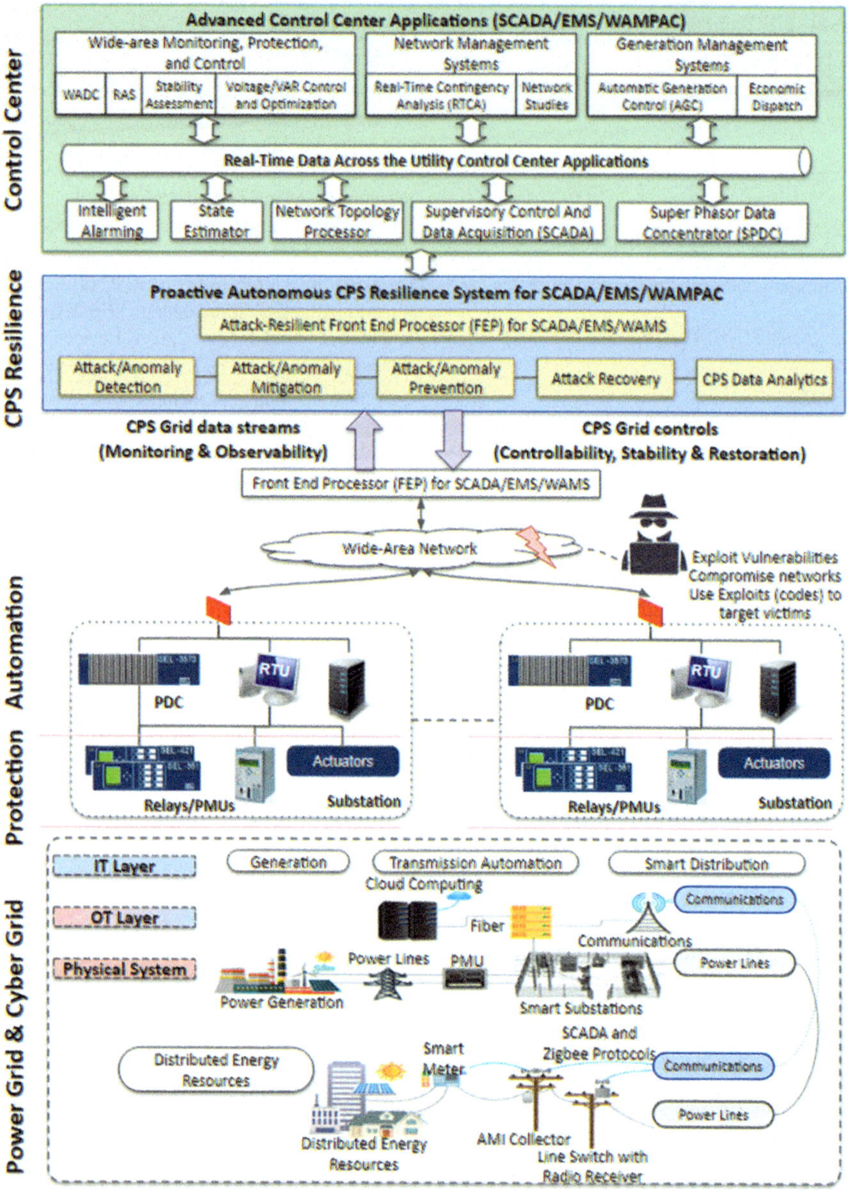

Fig. 8.2 Proposed CPS attack-resilient architecture for SCADA/EMS/WAMPAC systems

ADMS. Figure 8.3 shows both the designs with switching combinations. The *S1* enables Generic-ADMS mode and both the *S1* and *S2* enables Application-Specific ADMS mode.

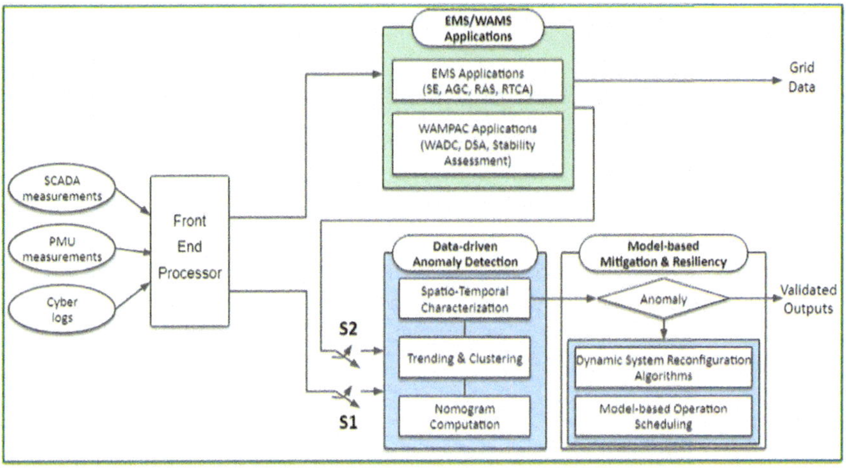

Fig. 8.3 Proposed anomaly detection and mitigation system for SCADA and WAMPAC

(1) Generic-ADMS model: In this model, we design the anomaly detection algorithms to detect anomalies on the raw grid data such as SCADA data streams and PMU data streams, and to the cyber-infrastructure data such as cyber-logs, intrusion detection logs, intrusion prevention logs, and Security Information and Event Management (SIEM) data logs. Subsequently, the anomaly mitigation algorithms provide decision signals such as optimal CPS grid topologies to the system operator to improve power grid resiliency.

(2) Application-Specific-ADMS model: In this model, we design anomaly detection algorithms to detect anomalies on the application-specific data streams. For example, the detection of false data injection on state estimation, data integrity attacks on wide-area applications such as AGC, RTCA, RAS, and WADC. Subsequently, anomaly mitigation algorithms provide application-specific mitigation strategies to the system operator to improve power grid resiliency tie-up with the corresponding application.

As shown in Fig. 8.2, we consider the Generic-ADMS model to build CPS resilience to the control center applications, namely the proactive autonomous resilience system (CPS-PARS). It receives CPS grid data streams to monitor and observe power system dynamics and issues CPS grid controls to the grid for controllability, stability, and restoration. As the CPS-PARS incorporates latency in the end-to-end framework, we consider designing different levels of serializing the CPS grid data streams to detect anomalies and provide the data to the upstream control center applications with minimum latency via attack-resilient FEP.

8.3.2 Control Center Network Modeling

The proposed attack-resilient framework comprises the following key activities:

(i) Modeling and analysis of the connectivity/topology of control center networks,
(ii) Modeling the interactions among various applications at control center network (SCADA, EMS, and WAMPAC) and analysis of their security and information flow properties (confidentiality, integrity, availability, latency), and
(iii) Identify functional system requirements and appropriate data sources for the proposed detection and mitigation within control center networks based on network communication rates, number of participating utilities, and accuracy of requirements.

8.3.3 Cyber-Physical Control Model

Most of the grid's current communication and control architecture is hierarchical, which has been very useful and well-structured. However, a growing number of protection and control applications rely on peer-to-peer interactions among controllers within a geographical region to achieve timely, collaborative decision-making. In this context, the agent-based peer-to-peer communication paradigm has been gaining importance in the modern grid environment. These principles can be used to develop a control center cyber-physical control model for achieving security and resiliency. In particular, the model identifies:

(i) Which nodes participate in the control loop,
(ii) What types of information being exchanged (attack vectors and defense measures), and
(iii) What is the rate of data in the control center cyber-physical control loop.

The anomaly detection incorporates multiple layers of intrusion/anomaly detection logic within each node, namely IT layer IDS, specification-based IDS for SCADA/WAMS protocol layer, and application-focused anomaly detection using machine learning that leverages cyber-physical system properties and application-specific parameters/features. The model-based mitigation algorithms leverage data from multiple sources such as cyber- and physical-system data and load forecast data to build applications' models. The overall approach is that when the measurements are detected to be anomalous, the system/application will rely on model parameters for making decisions while ignoring the measurement data. Our research will demonstrate the efficacy of the proposed attack-resilient architecture for critical control center applications, such as Automatic Generation Control (AGC)—an EMS application and Wide-Area Damping Control—a WAMPAC application.

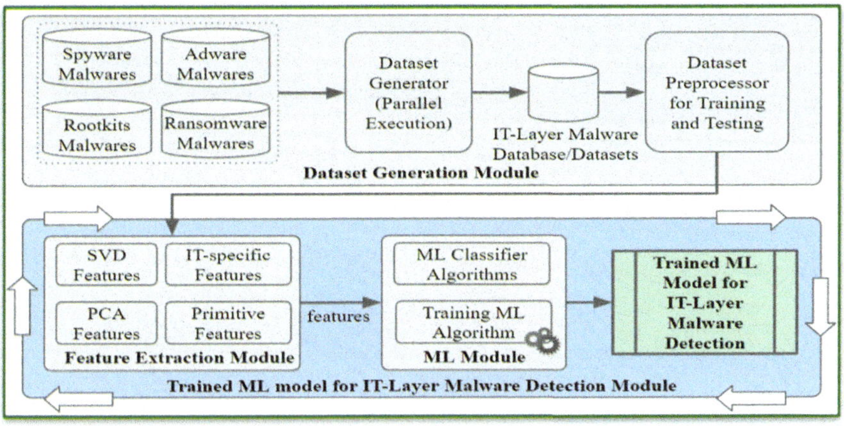

Fig. 8.4 IT anomaly detection

8.3.4 A Multi-layer Intrusion and Anomaly Detection System

Figures 8.4 and 8.5 show the schematic of the proposed multi-layer, defense-in-depth Anomaly Detection System (ADS) that will be deployed at the Utility Control Center and RTO/ISO Control Center for both IT and OT environments. The proposed ADS utilizes cyber-system properties, protocol properties, and application-specific properties and will be architected into three layers, as discussed below. The incoming data/signal/command/control from the network interfaces go through layer 1 IDS (IT-based) and then appropriate layer 2 IDS (protocol-specific), and finally, the suitable layer 3 ADS (application-specific) dependent on the protocol and application for which they belong. Since each layer detects (and filters, if needed) specific intrusions/anomalies of that layer, the overall ADS architecture is modular, enabling the integration of IDS/ADS for newer protocols/ applications. While there is a significant effort involved in developing such an integrated ADS, the project's research component primarily lies with application-specific ADS using machine learning while leveraging the models/techniques/platforms that exist for the other two layers.

8.3.4.1 Intrusion Detection System (IDS)

(1) IT-based IDS This layer focuses on IT-specific anomalies and signature-based detection on wide-area communication networks. The platform will leverage existing networks-based IDS tools, such as Snort, and Suricata, to identify traditional network-based attacks.

(2) Protocol specification-based IDS This layer focuses on building protocol-specific IDS for three major protocols that are commonly in use in the control center grid environment, namely DNP3, C37.118, and ICCP. The IDS incorporates each

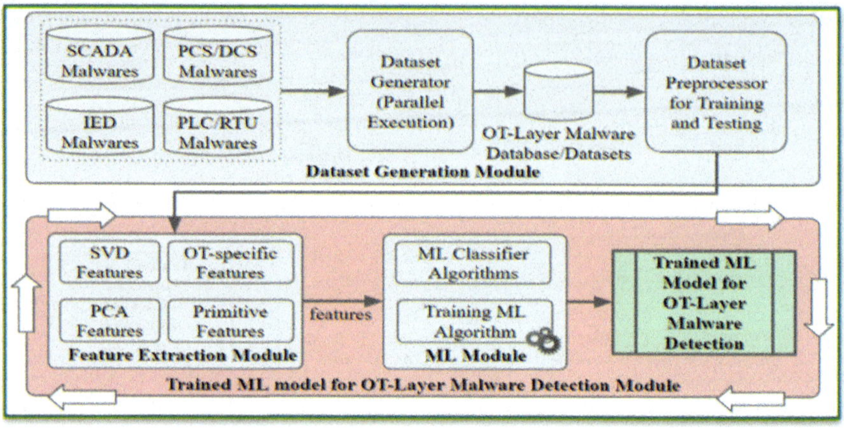

Fig. 8.5 OT anomaly detection

of these protocols' specifications to build a state-machine model that captures critical properties, including valid transitions, the right sequence of the transitions, and time-duration between the transitions. Various IDS tools such as Snort, Suricata, and Bro/Zeek were developed for IT environments. We consider leveraging the existing models/platforms to build IDS for this layer.

8.3.4.2 Application-Level ADS Using Machine Learning for both IT Data and OT Data

This layer focuses on building anomaly detection based on application layer properties. Machine learning provides a sound analytical framework to correlate data from multiple sources to create a baseline profile of the system's nominal behavior through a supervised or unsupervised learning process. Then, it uses classification algorithms to classify online events/measurements into normal or abnormal.

Figures 8.4 and 8.5 show the overall workflow of the proposed machine learning-based anomaly detection system for both the IT-data and OT-data, respectively. It can be summarized as follows:

(i) The approach begins with extracting sampled raw data from network packets based on various grid protocols preprocessed based on their associated grid applications (e.g., WADC).
(ii) Subsequently, we serialize the data and compute metrics, followed by preparing a vector of data metrics.
(iii) Each data vector will be fed into the online anomaly detection system to detect anomalies.

Either supervised or semi-supervised clustering machine learning algorithms would be a better fit to train the ML model offline, followed by updating the online ML models.

Traditional anomaly detection algorithms have two main shortcomings: (i) Most of the anomaly detector algorithms are unsupervised. It poses challenges to tune the parameters for enhancing the detection accuracy, and (ii) there are few principled algorithms to handle streaming data sets. The design includes active learning algorithms, which enable human operators to tune anomaly detector ensembles for both batch and streaming data settings. In the former settings, the entire data is available as a single instance at the outset, whereas, in the latter setting, the data streams are available as it arrives. Combining these approaches addresses the significant quantities of data communicated across an ISO network between a large number of potential utilities. Furthermore, multi-view isolation forest techniques enable the ability to explore causal dependencies between various power grid applications.

8.3.5 Physics Model-Based Mitigation for WAMPAC Applications

Figure 8.6 shows the schematic of a physics model-based mitigation approach integrated with a data-driven anomaly detection system for control center applications. The core of model-based mitigation is to establish an accurate model of the system for a given application. The model-based mitigation leverages data from multiple collaborative sources, such as cyber- and physical-system data, and/or load forecast to build a model of the applications. The overall approach is that when the measurement-based control or decision is detected to be anomalous, the system/application will

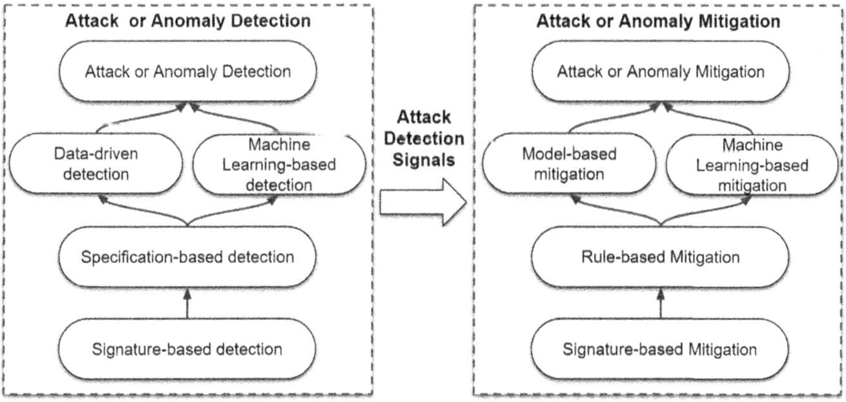

Fig. 8.6 Data-driven and model-based anomaly detection and mitigation system

rely on model parameters for making decisions ignoring the measurement-based decisions.

In the proposed framework, we achieve attack resilience by leveraging the models developed as a part of the data-driven anomaly detection approaches introduced in the previous section. Specifically, these would include leveraging a combination of spatio-temporal characterization of SCADA signals, trending and clustering of application-specific indicators, and computation of nomograms that provide bounds on operational limits for different system conditions during anomalies. These techniques ingest data from various information sources, including load and generation forecast data, statistical and historian data, load and generation schedule data, and system topology for model development. The mitigation module launches mitigation strategies such as dynamic reconfiguration of network actuators or power system typologies to minimize cyberattacks' effects on grid reliability, security, stability, and resiliency. Examples of such algorithms could be a dynamic reconfiguration of grid and SCADA systems, model-based operation scheduling. While the proposed mitigation techniques would cause suboptimal system performance, they would be designed to ensure minimal impacts on system reliability, stability, and, most importantly, they would not impede critical energy delivery functions.

8.3.6 Deep Learning and Machine Learning for Anomaly Detection

Machine Learning and Deep Learning algorithms realize Artificial Intelligence (AI) to real-world applications. The research on Deep Learning (DL) algorithms is widespread in self-driving cars, Image Recognition, and Voice Search and Voice-Activated Assistants [20]. A few works on the grid applications of Energy Market Price Forecasting and PMU data have commenced and are at the nascent phase.

In essence, similar to the ML, the DL includes statistical learning that extracts required features or attributes from raw data streams. The DL algorithms employ neural networks consisting of multi-layer hidden layers, massive datasets, and high-performance computational resources. Both ML and DL algorithms handle high voluminous datasets; however, ML algorithms require sophisticated techniques to process the datasets and extract features. It shows better performance for relatively smaller datasets. For example, if one has only a small set of data points, say 250, the k-nearest neighbors and decision trees, and other supervised and unsupervised ML algorithms would be much more valuable than fitting them into a DL model. One significant advantage of using DL over ML would be auto-encoders to construct the patterns and features to train models. As a trade-off, the DL requires extensive and substantial datasets for the training. Vendors like Google and Matlab use High-Performance Computing (HPC) to build rich datasets and make it available for various DL-based applications. As the DL uses a relatively huge dataset compared to ML and trains over many hidden layers with a robust computation, based on our theoretical

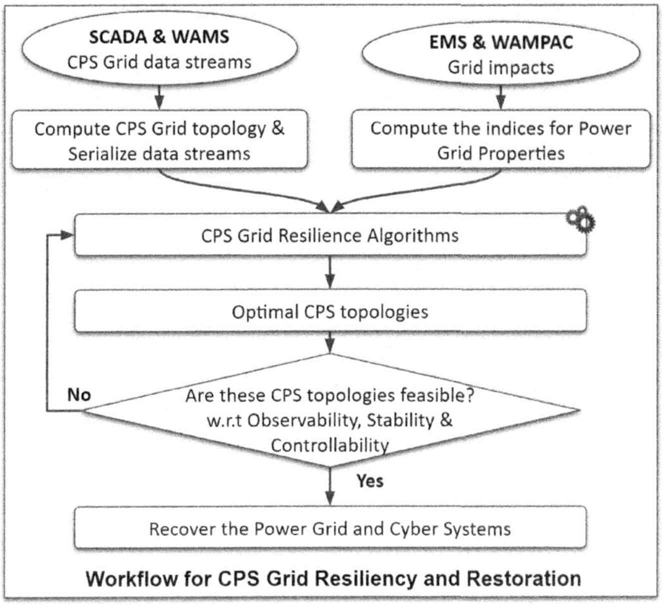

Fig. 8.7 Proposed CPS grid resiliency and restoration

knowledge, we believe that DL would provide better detection accuracy compared to ML. On the summary note, as a follow-up to our future research, we consider developing DL models and DL techniques to WAMS-based applications, anomaly detection, and anomaly mitigation applications.

8.3.7 CPS Grid Resiliency and Restoration

Figure 8.7 shows a schematic for CPS grid resiliency and restoration. It serializes the SCADA and WAMS grid data streams and compute CPS grid topologies. The CPS grid topologies include power system network topology interfaced with both sensors and actuators, and a logical cyber-system network topology interfaced with IEDs, RTUs, firewalls, IDS, Intrusion Prevention System (IPS), and Security Information and Event Management (SIEM) tools. On the other hand, the algorithm receives data from EMS and WAMPAC application to compute indices for the power grid properties such as observability, controllability, and stability. The CPS grid resiliency algorithm uses both the index and CPS grid topology to yield optimal CPS topologies. The algorithm validates the feasibility of these CPS topologies. Subsequently, it dispatches control signals and setpoints to both power grid actuators upon the feasible topology design, whereas it sends configuration signals, including parameter settings, to cyber-system devices.

8.4 HIL CPS Security Smart Grid Testbed

With growing cyber-threats to the power grid infrastructure, the need for building and hosting a realistic CPS testbed infrastructure for a wide range of use-cases ranging from R&D to education to training to attack-defense exercise has also grown significantly. The term *testbed* covers a broad spectrum: simulation-based, Hardware-in-the-Loop (HIL) testbed, Power Hardware-in-the-Loop (PHIL), test range deployments, and real-world industry deployments. It is not possible to do cyber-attack-defense experiments on real systems (and to some extent on test ranges as well). Moreover, it becomes prohibitively expensive to replicate real systems for security and performance evaluations. An alternative is to perform large-scale computer-based simulations, but that involves compromising the modeling accuracy. In this respect, HIL and PHIL testbeds provide a pragmatic balance between the low cost and ability to accurately capture real system characteristics. These testbeds are also useful as they provide a methodology to study the complex cyber-physical interactions in the power grid, which cannot be accurately modeled using traditional modeling and simulation tools. We developed a hardware-in-the-loop CPS security smart grid testbed architecture as shown in Fig. 8.8 at Iowa State University. The recent R&D using the CPS testbed can be found in [1, 17, 21–26, 28, 32]. The description of the modules is as follows.

- Control Center module: It includes the Siemens PowerTG software to establish SCADA functionality. We model the desired power system models in PowerTG after preparing a database for the physical IEDs using the Substation Remote Terminal Units (RTUs). We configure the PowerTG to establish the SCADA communication with the substation RTUs through the DNP3 protocol.
- Substation RTU module: It includes the Siemens SICAM PAS software module to establish communication between the Control Center module and the physical IEDs. We deploy customized Configured IED Description (cCID) files in the substation RTUs for the configuration of data objects in the RTU derived from the logical nodes of the physical IEDs. The substation RTU serializes the DNP3 packets coming from the control center and transmits them to the IEDs based on a telegram address assigned to each data object using IEC-61850 MMS protocol. We have configured a total of 32 data objects (also referred to as remote bits) ranging from SPCSO01 to SPCSO32 in each substation RTU.
- Physical IEDs Configuration module: It consists of physical IEDs such as SEL-351 and SEL-421 devices and their respective configuration tools. These IEDs are mapped to the SICAM PAS substation RTU and the power system models designed in the OPAL-RT. We use IEC-61850 MMS protocol for the communication between the IEDs and SICAM PAS whereas IEC-61850 GOOSE protocol for the interaction between the IEDs and OPAL-RT power system models. The cCID file for each device is prepared and deployed in the respective physical IED using IED configuration tools over File Transfer Protocol (FTP). The cCID files define the data objects that need to be transmitted to the OPAL-RT module.

- OPAL-RT Power System Model (PSM) module: It includes the power system models designed and simulated in the OPAL-RT Real-Time Digital Simulator (RTDS). We map the desired physical IEDs to the OPAL-RT through the IEC-61850 Input/Output (I/O) interface of the RT-LAB software. These I/Os are mapped to the circuit breaker blocks modeled in the power system models. The I/O interfaces receive GOOSE packets with the trip/close status information from the physical IEDs which themselves are triggered by trip/close commands coming from the control center through the substation RTU.
- Cyber-System module: It includes the configuration of physical switches and the VMWare Virtual Local Area Networks (VLANs) between the physical device modules and software modules. We consider the VMWare VLANs for the establishment of the wide-area network between SCADA, substation RTUs, and WAMPACS Applications.
- WAMPACS Applications module: It consists of software PDCs of the OpenPDC software system and hardware SEL-3573 PDCs. We configure the PDCs to the virtual PMUs that are modeled inside the OPAL-RT power system models. It uses the IEEE C37.118-2011 protocol for the synchrophasor communication between hardware/software PDCs and the virtual PMUs in OPAL-RT.

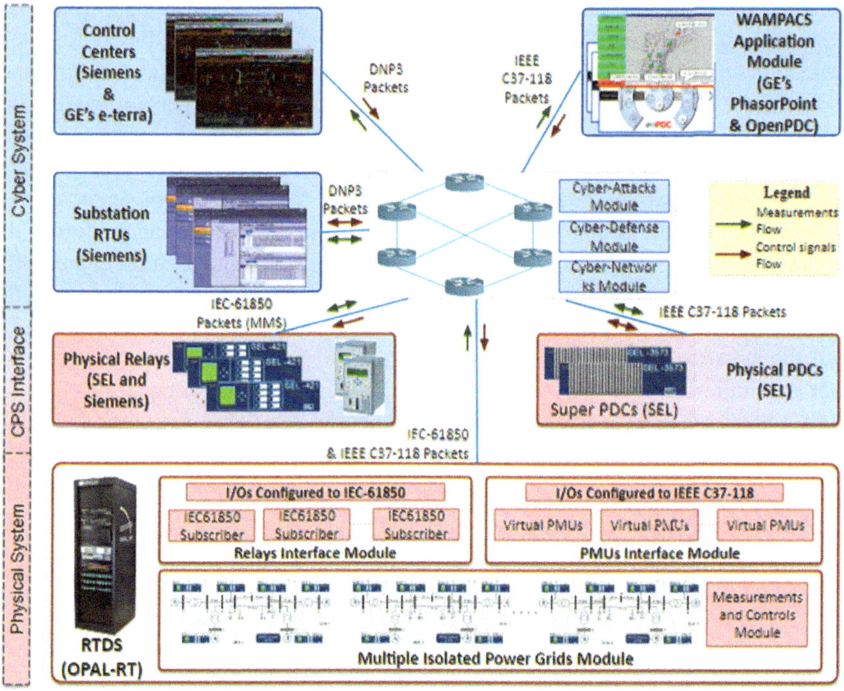

Fig. 8.8 Hardware-in-the-loop CPS security smart grid testbed at Iowa state university

8.5 Case Study: Attack-Resilient WADC Using Machine Learning

WADC is one of the critical Wide-Area Control Systems (WACS) applications in modern power systems. A wide-area damping controller, typically located at the control center, synthesizes a set of remote measurement signals to generate WADC signals to dispatch to the geographically distributed actuators. Stealthy and coordinated cyberattack vectors may bypass conventional cybersecurity measures to disrupt WACS applications' seamless operation. Adversaries deploy these cyberattack vectors to modify the compromised data channels (either measurement signals, control signals, or both), which manifests the cyberattack anomalies to the actual data WACS applications. We recently presented an Anomaly Detection and Mitigation (ADM) for WADC using machine learning in [21]. We present here an expanded discussion of the research work.

As an example, one of our approaches for detecting attacks in WADC using Machine Learning, shown in Fig. 8.9, relies on synchrophasor measurement data and the WADC signals. The geographically distributed PMUs compute the synchrophasor measurements from the Current Transformers (CTs) and Potential Transformers (PTs) and transfer them to the configured local and super Phasor Data Concentrators (PDCs). The WADC module processes the selected synchrophasor data from PDC and synthesizes a control signal. The proposed machine learning-based Anomaly Detection and Mitigation (ADM) module processes all the data signals from super PDC and the WADC signal to detect cyberattack anomalies and executes model-based mitigation if it detects any cyberattack anomalies. Figure 8.10 shows the detection of anomalies in the current cycle when the cyberattacks are present in measurement signals and the mitigation strategy combined with the WADC tool sends a tuned control signal to the actuators. Figure 8.11 shows the detection of anomalies in the next cycle when the cyberattacks are present in control signals. Subsequently, the mitigation strategy combined with the WADC tool sends a tuned control signal to the actuators. However, as the cyberattacks occur between the control center application and the actuator, the actuator receives a tuned control signal along with cyberattack anomalies. In this operation mode, we consider two mitigation strategies—(1) Only-PSS mode and (2) Reconfiguration mode. The details are discussed in [21].

8.5.1 Machine Learning-Based Anomaly Detection for WADC

Machine learning is a well-known technique that uses data and produces a program to perform a task. For example, consider an activity involving measurement signals (from phasor measurement units) and control signals (output of WADC) to detect attack activities such as normal, perturbation, attack, or attack-and-perturbation. We can detect attack activity, either by a standard mathematical-model-based approach or

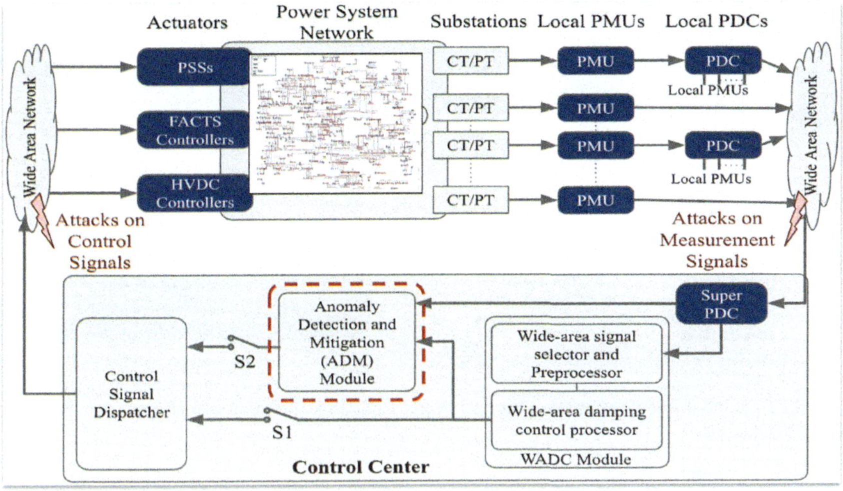

Fig. 8.9 A CPS control loop architecture for WADC application

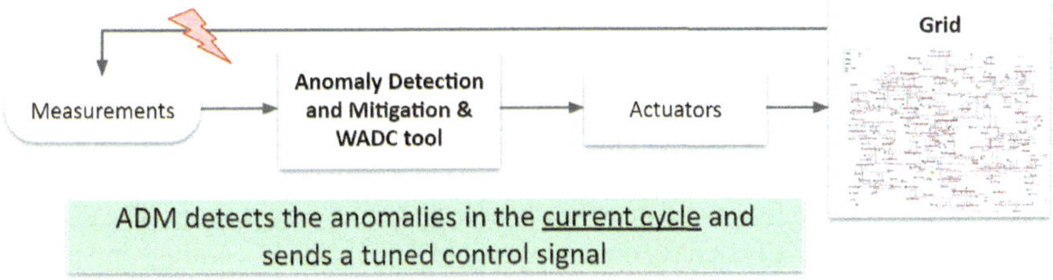

Fig. 8.10 Cyberattacks on measurement signals

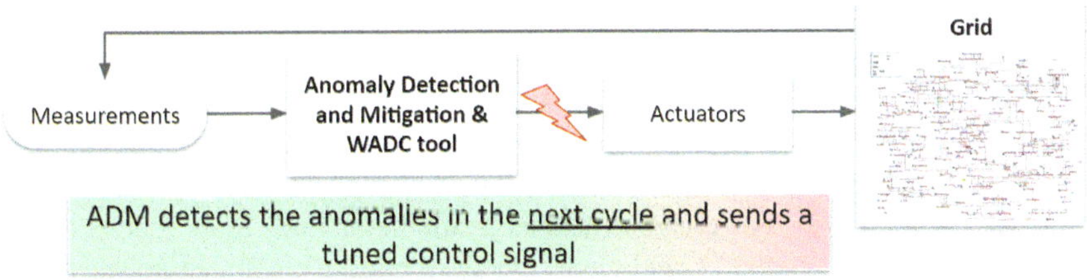

Fig. 8.11 Cyberattacks on control signals

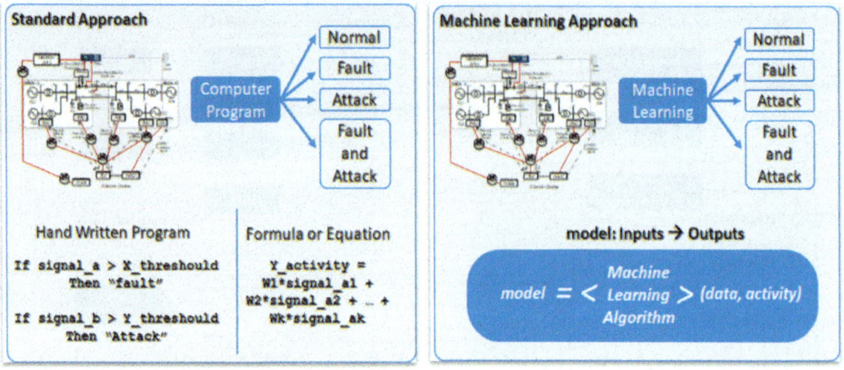

Fig. 8.12 Standard programmatic approach versus machine learning-based approach for attack activity detection

by using machine learning. Figure 8.12 shows both approaches for an attack activity detection. In the first approach, we devise mathematical models to build an algorithm that detects activity type. On the other hand, we provide the Machine Learning algorithm with input and output data sets and learn the program to solve the attack activity. This process is referred to as training, and the output is a model that can now be used to detect activity from new measurement and control signals. We do not explicitly mention a specific Machine Learning algorithm because there are plenty of algorithms to choose from, and each has its strengths and weaknesses. The essential requirement is if you do not have data, you cannot do Machine Learning, and it needs to be considered when the task is complex, and there is no mathematical model or system of equations to solve it.

We use supervised ML algorithms, namely Support Vector Machine (SVM), Decision Trees (DT), k-Nearest Neighbor (KNN), Naive Bayes, discriminant analysis, logistic regression, and neural networks for the classification of data. The SVM and KNN exhibit more excellent performance for the classification of anomalies; on the other hand, the KNN and DT converge more quickly and shorten the training time. Hence, we consider DT and KNN Machine Learning algorithms for anomaly detection. We configure the PowerCyber testbed available at Iowa State University for a high-performance computational environment to build a combined power system operating conditions and cyberattack events-based dataset generation model using a parallel execution approach that can be used for any large-scale power grid models.

8.5.2 *Results and Discussion*

We have considered both open-loop and closed-loop operation of integrated ML-based and model-based ADM to evaluate ML-based approaches' efficacy.

(1) Open-loop operation: In this operation, we conduct training followed by testing ML algorithms to determine the True-positive rate, True-negative rate, False-positive rate, and False-negative rate. Subsequently, we determine the following Key Performance Indicators (KPIs) [29] to measure ML algorithms' efficacy.

$$Accuracy = P_A = \frac{\alpha - \beta}{\alpha} \tag{8.1a}$$

$$Reliability = P_R = \frac{\alpha_u - \beta_{ds}}{\alpha_u} \tag{8.1b}$$

$$Security = P_S = \frac{\alpha_s - \beta_{du}}{\alpha_s} \tag{8.1c}$$

$$MisDetection = P_M = \frac{\beta_{ds}}{\alpha} \tag{8.1d}$$

$$FalseAlarm = P_F = \frac{\beta_{du}}{\alpha} \tag{8.1e}$$

$$Correctness = P_C = 1 - \frac{6\sum d_i^2}{\alpha(\alpha^2 - 1)} \tag{8.1f}$$

Here, the notation is given as follows:

- α_u: Number of attack cases (both *attack* type and *FaultAndAttack* type),
- α_s: Number of cases without attacks (both *Normal* and *Fault* types),
- α: Number of cases where $\alpha_u + \alpha_s$,
- β_{ds}: Number of cases detected as normal or fault cases,
- β_{du}: Number of cases detected as attack or fault-and-attack,
- $\beta = \beta_{ds} + \beta_{du}$: Number of mis-classification cases, and
- $d_i = x_i - y_i$, where x_i denotes the number of attack cases injected, whereas y_i denotes the number of attack cases detected (i.e., β).

Figure 8.13 shows True Positive and False Negative rates for the trained ML models with domain-specific features. We observed that the trained ML model with domain-specific features had exhibited an enhanced detection accuracy at all the activities. The trained KNN ML model (Fig. 8.13) has shown better performance with 96% accuracy. The model predicts less than 1% of *Attack* activity as *Normal* and 3% of *Attack* activity as *PerturbationAndAttack* activity. The model predicts the *PerturbationAndAttack* as *Perturbation* with 1%, as *Attack* with 5%, and as *PerturbationAndAttack* with 94% accuracy. The characteristic patterns of *Attack* and *PerturbationAndAttack* can be in proximity to *Perturbation* characteristics by stealthy cyberattack vectors. It results in a significant challenge to ML for activity detection with higher accuracy. The inclusion of domain features such as TKEO and WASMs provided better measures to train the ML model to detect the *PerturbationAndAttack* activity with 94% accuracy. It can be further improved by fine-tuned parameter settings of the ML models and dynamic window length.

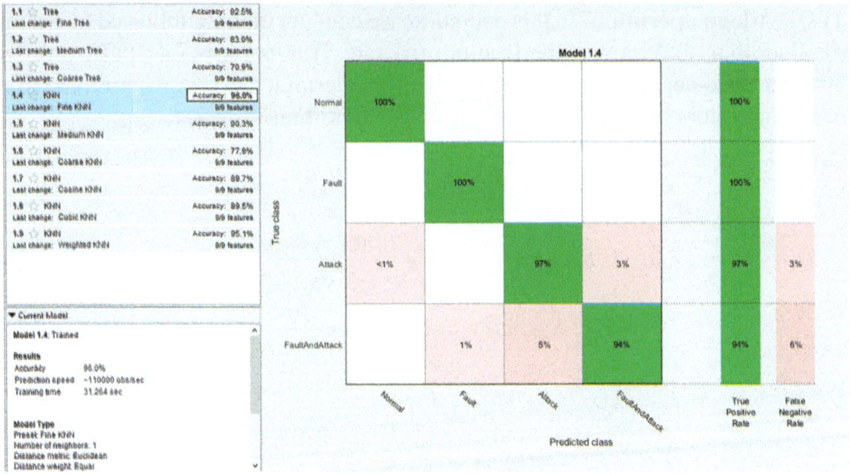

Fig. 8.13 Detection accuracy of trained ML model using KNN with proposed domain-specific features

(2) Closed-loop operation: In this operation, we obtain time-domain characteristics for both with ADM and without ADM against various stealthy data integrity attacks on measurement signals and control signals. Figures 8.11 and 8.12 presented in our ADM for WADC paper [21] report the combined performance of ADM and WADC tool operation in Hardware-in-the-Loop (HIL) close loop environment for data integrity attack vectors such as pulse attacks and ramp attacks on measurement signals and control signals.

8.6 Conclusion

In this chapter, we proposed an ML-based and Physics-based attack-resilient framework for WAMPAC. Subsequently, we discussed the anomaly detection and mitigation system designs for SCADA and WAMPAC control center applications. We discussed attack surface analysis for CPS control loop applications and identified potential cyberattacks vectors such as DoS, Data integrity, Replay, and Timing attacks on measurement signal channels and control signal channels. We provided a case study to illustrate the attack-resiliency using machine learning for time-critical wide-area control system applications such as WADC. The ADM for WADC using a machine learning case study considered the physics-based and signal-entropy-based feature extraction to increase the trained ML model's accuracy and robustness. We discussed the mathematical models and algorithms in [21]. Our future research includes the design and implementation of ML and DL algorithms for defense-in-depth CPS security for other WAMPAC applications and DERMS applications.

Acknowledgements This research is funded in part by US NSF Grant # CNS 1446831, and US DOE Grant # DE-OE0000830.

References

1. Ashok, A., Govindarasu, M., Wang, J.: Cyber-physical attack-resilient wide-area monitoring, protection, and control for the power grid. Proc. IEEE **105**(7), 1389–1407 (2017)
2. Baker, S.A., Waterman, S., Ivanov, G.: In the crossfire: critical infrastructure in the age of cyber war. McAfee, Inc. (2009)
3. Boyer, S.A.: SCADA: Supervisory Control and Data Acquisition. International Society of Automation (2009)
4. Cauley, G., Lauby, M.: High-impact low-frequency event risk to the North American bulk power system. NERC, Atlanta, GA, USA, Technical Report (2010)
5. Chakrabarti, S., Kyriakides, E., Bi, T., Cai, D., Terzija, V.: Measurements get together. IEEE Power Energy Mag. **7**(1), 41–49 (2008)
6. Cybersecurity and Infrastructure Security Agency, US Department of Homeland Security: Russian government cyber activity targeting energy and other critical infrastructure sectors. Alert (TA18-074A) (2018)
7. Dragos, Inc.: Crashoverride: Analyzing the Threat to Electric Grid Operations. (2017) [Online]. https://dragos.com/blog/crashoverride/CrashOverride-01.pdf
8. Energy Sector Control Systems Working Group: Roadmap to achieve energy delivery systems cybersecurity. Energetics, Inc. [Online] (2011). https://www.energy.gov/sites/default/files/Energy%20Delivery%20Systems%20Cybersecurity%20Roadmap_finalweb.pdf
9. Falliere, N., Murchu, L.O., Chien, E.: W32. Stuxnet dossier. White paper, Symantec Corp., Secur. Response **5**(6), 29 (2011)
10. Folkerth, L.: Forensic analysis of industrial control systems. SANS Institute InfoSec Reading Room (2015)
11. Hawk, C., Kaushiva, A.: Cybersecurity and the smarter grid. Electr. J. **27**(8), 84–95 (2014)
12. Idaho National Laboratory: Vulnerability analysis of energy delivery control systems. INL/EXT-10-18381 (2011)
13. Kuipers, D.: Common cyber security vulnerabilities observed in control system assessments by the inl nstb program. Idaho National Lab.(INL), Idaho Falls, ID (United States), Technical Report (2008)
14. Lamp, J., Rubio-Medrano, C.E., Zhao, Z., Ahn, G.J.: Exsol: Collaboratively assessing cybersecurity risks for protecting energy delivery systems. In: Proceedings of 7th Workshop on Modeling and Simulation of Cyber-Physical Energy Systems (MSCPES), pp. 1–6. IEEE (2019)
15. Lee, M., Assante, M.J., Conway, T.: TLP: White-Analysis of the Cyber Attack on the Ukrainian Power Grid-Defense Use Case (2016)
16. McDonald, J.D.: Electric Power Substations Engineering. CRC Press (2016)
17. Mohan, S.N., Ravikumar, G., Govindarasu, M.: Distributed intrusion detection system using semantic-based rules for SCADA in smart grid. In: Proceedings of IEEE/PES Transmission and Distribution, Conference and Exposition (T&D) (2020)
18. Nussbaum, D., Pickl, S., Dupuy, A., Nistor, M.S.: The nexus between cyber security and energy security. In: Proceedings of European Conference on Cyber Warfare and Security, pp. 228–235 (2016)
19. Qassim, Q.S., Jamil, N., Daud, M., Patel, A., Ja'affar, N.: A review of security assessment methodologies in industrial control systems. Inf. Comput. Secur. **27**(1), 47–61 (2019)
20. Rani, K.S., Kumari, M., Singh, V.B., Sharma, M.: Deep learning with big data: an emerging trend. In: Proceedings of 19th International Conference on Computational Science and Its Applications (ICCSA), pp. 93–101 (2019)

21. Ravikumar, G., Govindarasu, M.: Anomaly detection and mitigation for wide-area damping control using machine learning. IEEE Trans. Smart Grid (2020)
22. Ravikumar, G., Hyder, B., Govindarasu, M.: Efficient modeling of HIL multi-grid system for scalability & concurrency in CPS security testbed. In: Proceedings of North American Power Symposium (NAPS) (2019)
23. Ravikumar, G., Hyder, B., Govindarasu, M.: Efficient modeling of IEC-61850 logical nodes in IEDs for scalability in CPS security testbed. In: Proceedings of IEEE/PES Transmission and Distribution Conference and Exposition (T&D) (2020)
24. Ravikumar, G., Hyder, B., Govindarasu, M.: Hardware-in-the-loop CPS security architecture for der monitoring and control applications. In: Proceedings of IEEE Texas Power and Energy Conference (TPEC), pp. 1–5 (2020)
25. Ravikumar, G., Hyder, B., Govindarasu, M.: Next-generation CPS testbed-based grid exercise-synthetic grid, attack, and defense modeling. In: Proceedings of Resilience Week (RWS), pp. 92–98 (2020)
26. Ravikumar, G., Nicklaus, A., Govindarasu, M.: Cyber-physical smart light control system integration with smart grid using zigbee. In: Proceedings of IEEE Power Energy Society Innovative Smart Grid Technologies Conference (ISGT), pp. 1–5 (2020)
27. Ravikumar, G., Ramya, G., Misra, S., Brahma, S., Khaparde, S.A.: iPaCS: an integrative power and cyber systems co-simulation framework for smart grid. In: Proceedings of IEEE Power and Energy Society General Meeting, pp. 1–5 (2017)
28. Ravikumar, G., Singh, A., Babu, J.R., Moataz A, A., Govindarasu, M.: D-IDS for cyber-physical der modbus system - architecture, modeling, testbed-based evaluation. In: Proceedings of Resilience Week (RWS), pp. 153–159 (2020)
29. Ravikumar, G., Khaparde, S.A.: Taxonomy of pmu data based catastrophic indicators for power system stability assessment. IEEE Syst. J. **12**(1), 452–464 (2016)
30. Shelar, D., Sun, P., Amin, S., Zonouz, S.: Compromising security of economic dispatch in power system operations. In: Proceedings of 47th Annual IEEE/IFIP International Conference on Dependable Systems and Networks (DSN), pp. 531–542. IEEE (2017)
31. Sridhar, S., Govindarasu, M.: Model-based attack detection and mitigation for automatic generation control. IEEE Trans. Smart Grid **5**(2), 580–591 (2014)
32. Sridhar, S., Hahn, A., Govindarasu, M.: Cyber-physical system security for the electric power grid. Proc. IEEE **100**(1), 210–224 (2012)
33. Terzija, V.: Wide area monitoring protection and control-wampac. In: Proceedings of IET-UK International Conference on Information and Communication Technology in Electrical Sciences (ICTES 2007), pp. I–1. IET (2007)
34. Terzija, V., Valverde, G., Cai, D., Regulski, P., Madani, V., Fitch, J., Skok, S., Begovic, M.M., Phadke, A.: Wide-area monitoring, protection, and control of future electric power networks. Proc. IEEE **99**(1), 80–93 (2010)
35. The Smart Grid Interoperability Panel–Smart Grid Cybersecurity Committee: Nistir 7628: Guidelines for smart grid cyber security. Priv. Smart Grid **2** (2010)

Index

H. Ishii and Q. Zhu (eds.), *Security and Resilience of Control Systems*,
Lecture Notes in Control and Information Sciences 489,
https://doi.org/10.1007/978-3-030-83236-0

W

X

Z

The manufacturer's authorised representative in the EU is Springer Nature Customer Service Centre GmbH, Europaplatz 3, 69115 Heidelberg, Germany. If you have any concerns regarding our products, please contact ProductSafety@springernature.com

Printed and bound by CPI Group (UK) Ltd, Croydon, CR0 4YY
30/09/2026
02225445-0001